THE ROYAL NAV[illegible] AUXILIARY SERVICE

The first twenty-five years

JOHN MURRAY

MARITIME B[illegible]

Contents

First published in Great Britain by
Maritime Books, Liskeard, Cornwall.

ISBN 0 907771 37 8

Typeset and printed in Great Britain by
Penwell Ltd., Parkwood, Callington, Cornwall.

It is a great honour for me to have been invited to become the first Patron of the Royal Naval Auxiliary Service and to contribute this foreword to a book which tells us so much of its formation, character and development over the last 25 years.

I met for the first time many of the men and women who form this remarkable, little-known and largely unsung Service when I reviewed the RNXS both ashore and afloat at Whale Island in Portsmouth in their Silver Jubilee year.

I commend this history not only as a graphic record of their first quarter of a century but also as a tribute to those who give so much in contributing to the defence of their own country for little tangible reward and who by doing so give all of us hope for the future.

Michael

Preface

The Royal Naval Auxiliary Service (RNXS) is a uniquely British Institution, publicly funded but with limited resources and, under the guidance of a small staff, almost entirely dependent on voluntary effort. It has grown steadily in expertise and stature in response to the varied requirements of the Service which it supports whilst remaining modest and decentralised, and has become adept at improvisation and self-help.

Although the staff may not readily agree, it can be judged fortunate that the RNXS has had more demands placed upon it than it has been able comfortably to satisfy. History, ancient and recent, is replete with examples of voluntary organisations which, lacking clear objectives or with not enough to do to attain them, quickly lose cohesion or take up tasks for which they have no mandate. The support of public funds implies public accountability and careful control of expenditure but more importantly, it obliges the organisation thus supported to stay in tune with current thinking and be responsive to changes in the requirement.

The story of the organisation, of how it came about and how it was managed, is told in the following pages in chapters one to six and ten to twelve. But the heart of the book, quite properly, tells of the heart of the Service—the Units, their localities and their people. Thus chapters seven, eight and nine deal, respectively, with the Portsmouth, Plymouth and Scotland and Northern Ireland Areas and the rich complex of places and personalities without which the history would be lifeless.

In a small book there is not space to mention every volunteer although each deserves to be remembered. For their loyalty, dedication and patience. RNXS members, past and present, have earned the right to celebrate their Silver Jubilee and to enjoy the satisfaction of a job well done. The RNXS remains in fine fettle—stretched, helpful, friendly and efficient but by no means complacent. Its members are proud to be of service to their country.

Chapter One

Origins

The First Minewatchers (c.1952) pose for their photograph at HMS Vernon *(Portsmouth) . . . . from humble beginnings . . . . Uniforms came soon afterwards.*

The RNXS was formed, or more accurately reformed, from the Royal Naval Minewatching Service in 1962 in response to the nuclear threat to United Kingdom ports. Auxiliarymen were to be readily available to assist in the tasks of evacuating major ports and despatching larger and faster merchant vessels overseas. The remaining ships, together with important port plant, were to be dispersed to safe anchorages along the coasts or in the islands. The sailing of merchantmen overseas involved setting up assembly anchorages where ships might be formed into convoys, a naval escort or surveillance aranged and information provided on routeing, intelligence and communications. Auxiliarymen were required at Assembly anchorage control. These tasks were not new ones, indeed they had been performed by naval and merchant naval personnel throughout the past centuries. The nuclear threat, however, had now cut the time available for transition from peace to war dramatically and the necessary organisation had to be developed or adapted to cope.

Before explaining more fully the RNXS role it is appropriate to note that the origins of such voluntary service lie deep in our history; geography, too, plays a part. The United Kingdom is fortunate in being well placed to control the sea routes to and from Northern Europe and in having so many excellent ports spread all round her long coastline. The defence of the ports was always a joint responsibility of the Army and Navy, with the Navy providing seaward defence and the Army manning the shore guns and forts; the Martello Towers, for example. Logistic support for ships was provided by civilians in the employ of the Navy Board at the Royal Yards or commercially at other ports under the direction of shipowners or their agents. Although in war more comprehensive arrangements evolved for commercial ports in the front line, it was not until the end of the 18th century, during the Napoleonic Wars, that the Admiralty decided to establish a Corps of Sea Fencibles with Companies at ports and rivers to defend them and to support naval ships using them. Volunteers willing to train and to be available when called on for duty were paid a retainer of one shilling a month and provided with a uniform. The Fencibles were drawn from all types;

Two of the earliest pieces of equipment. A basic minewatching sight—and the faithful duffle-coat!

tradesmen, builders and fishermen to name a few, their officers being local gentry or Service officers on half pay. The companies were located chiefly in the south from South Wales to the Wash. The Sea Fencibles provided many valuable services, for example in provisioning naval ships and taking charge of captured vessels. They manned local craft carrying guns to defend their ports and provided a degree of deterrence when there was for some years a real threat of invasion. It is of interest that the manning of the fleet during the Napoleonic Wars was a persistent problem and the press gangs were hept busy; local fishermen, unless members of the Sea Fencibles, were not excused! The end of the war in 1815 virtually eliminated the threat of invasion and the Sea Fencibles began to disband. Some companies however, at the Cinque Ports for example, continued in being for up to twenty years, albeit on a de-activated basis.

During the next hundred years, while considerable attention was paid to the defence of naval bases and the Royal Yards in the south, little thought was given to the defence of ports generally or to the merchantmen in them. Britain had a powerful navy to counter threats of invasion other than on the Channel coasts where the short sea crossing gave little warning and the advantage, to a powerful enemy, of surprise. The change from sail to steam and advances in weaponry necessitated more demanding training and increased professionalism; the naval and marines reserves were formed, it being no longer practicable to rely on the Press Gangs. When the challenge from Germany came in 1914, the Royal Navy and its reserves were ready but the needs of commercial shipping had been ignored. U-boat warfare proved devastating and, although a Protection of Shipping Organisation was set up, it remained inadequate until convoying was introduced in 1916. Coastal defence outside the naval bases relied on the Coastguard and the police. Also in 1916, the Admiralty set up an organisation for observing and reporting Zeppelins in the south-eastern corner of the country, involving the Coastguard, Police and many civilian volunteers. Zeppelins were tracked successfully both visually and aurally (but somewhat haphazardly) crossing the coast and approaching London, the reports being transmitted by telephone to a room in the Admiralty. This organisation, which became the Observer Corps, also undertook aeroplane spotting; it was disbanded in 1918 but re-instituted by the Air Ministry in 1925 using a dedicated permanent staff to organise and train a large body of specially recruited volunteers.

By 1918 the Naval Control of Shipping (NCS) Organisation was functioning well in the face of the U-boat threat, mines and sea raiders, and an integrated organisation consisting of reserves, pensioners and civilians had been built up to serve it. So in 1939, with the air threat to shipping added to the previous three continuing and intensifying threats, experience from the First World War enhanced by subsequent planning and development enabled early action to be taken to establish organisations for local defence and protection of shipping. Nevertheless, the transition from peace to war took up to six weeks in some port areas, while the personnel were found, appointed and drafted. Several losses from U-boat attack and mining occurred in the first week of the war. Equally important at this time were the self-inflicted casualties from collisions and groundings caused by the blackout (of shore lights and navigational aids), the establishment of war channels, lack of control and routeing and non-receipt of navigational warnings. Indeed during the first weeks of the war there were more merchant ships lost or damaged through these causes than by enemy action.

The local defence organisation in the Second World War covering coastal, seaward and harbour defence was manned by all three services using personnel of both sexes and of all grades and types of service, the majority being reserves, auxiliaries or pensioners. Simultaneously, together with the Naval Control of Shipping Organisations, a vast expansion in numbers and responsibilities mushroomed at home and overseas, in the front line and in the rear, to cater for the re-occupation of ports vacated by the enemy and to provide mobile or temporary bases for the next steps forward to victory. However, at the end of the war these Organisations were run down rapidly and the remnants returned to civilian and commercial control. Only in some areas where the danger of unswept mines remained, were active Naval Control authorities retained. The majority of the various arms of local defence, for example Boom Defence, Seaward Defence and Degaussing, were put on a care and maintenance basis, with their outline plans for re-activation and manning being overhauled from time to time by busy planning staffs preoccupied with other priorities. Gradually through the late 1940s, the 1950s and 1960s financial stringency and the increasing costs of upkeep or replacement of both headquarters and equipment caused many of these services to be reduced to no more than ideas on paper. The overwhelming threat and deterrent power of nuclear weapons appeared to render them insignificant.

Chapter Two

The Minewatchers and their metamorphosis into the RNXS

Just One Mine—easily laid—not easily found—but could, and still can, pose a serious threat around our coasts.

Immediately after the Second World War, with no serious naval threat remaining to the security of the United Kingdom but with a very real threat of national bankruptcy, the rundown of wartime defence structures was carried through as rapidly as possible. However, minewarfare was still much in the forefront of naval affairs with the vast clearance task of the minefields that had been laid by both sides. Before long, Soviet imperialist aspirations led to growing fears of a threat to Western security and in due course to the formation of the North Atlantic Treaty Organisation. The Russians were known to have large stocks of mines and to be continuing the development of a sophisticated arsenal. Since the Soviet Navy at that time was a large but predominately coastal force, it was considered that minelaying would be chiefly by air into the principal trade routes of Western Europe. After four years the primary clearance task had been completed and, prompted by Admiral Sir Arthur Power, Commander-in-Chief Channel, the Allied Navies began a thorough review of their minesweeping forces.

In planning to meet the new threat, a major consideration was deemed to lie in obtaining intelligence and warning of enemy mines, so that they could be avoided and minesweepers need spend the minimum time in widespread searching, the better to concentrate on clearing mines from essential channels. In addition to planning the construction of Coastal and Inshore Minesweepers, the Naval Staff initiated two complimentary moves, the first to provide a force of men and women volunteers to cover all vulnerable anchorages, bays and inshore channels with the task of reporting low flying aircraft and minedrops, working in conjunction with HM Coastguard and the Air Defence of the United Kingdom Organisation. The second move was to develop a high definition radar set capable of detecting mines dropped from aircraft and, most importantly, fixing the positions of the splashes as the mines entered the water. This project, codenamed AVOCADO, went ahead well and an experimental assembly underwent successful trials at Portland in 1952. Unfortunately, the cost of full development and production, at £80,000 per set, was prohibitive. The required high definition could only be achieved at the expense of range and it was estimated that some 300 sets would be needed for effective coverage; such expediture was out of the question and the project was put on ice.

Meanwhile, much difficulty was encountered throughout 1950-51 in trying to form this new reserve element for local defence purposes. The task was offered to the Admiralty Commanding Reserves (ACR), but was refused on the grounds that the job of shore lookouts did not fit with and would detract from his main task of providing a seagoing reserve. It has to said that there was a deal of war weariness about and most people were still busy adjusting to peace time, so ACR could be forgiven for expressing doubts over the numbers of volunteers likely to be forthcoming. Back in the Admiralty, the Board was determined that some action was essential to meet the mine threat and decided to change the emphasis from naval reserve to civilians in uniform for this force. The Secretary of the Admiralty (Sir John Lang) was instrumental in recommending that the constitution of the Royal Observer Corps be adopted as a model for this new body and Commander Philip Sharpe (later Rear Admiral) of the local defence section of Naval Plans Directorate started the task of formation. It was on 26th January 1952 that the Prime Minister, Sir Winston Churchill, announced the formation of the Naval Minewatching Service in the House of Commons.

The Mine . . .

Spot It

Report It

Record It

Plot It

The first step in the creation of this new Service was to appoint retired officers, as Civil Servants, to the staffs of the Commander-in-Chief at the major naval ports as Command Minewatching Officers and the process of forming Units at ports and anchorages throughout the United Kingdom began. The key to success lay in finding volunteers to act as Heads of Units willing to undertake the tasks necessary for recruitment of members and their training. HMS *Vernon* was put in overall charge of minewatching training and an active service officer, Lieutenant Commander L F Foley, was appointed for the task. Area Training Officers were then appointed to each Command to supervise the training of Units and Administrative Officers were appointed to some thirteen major United Kingdom ports to deal with the provision of equipment and clothing and to handle the allowances and expenses incurred by the volunteers.

Recruitment surpassed expectations, particularly on the South and East Coasts where the threat was greatest. Despite a lack of facilities, the enthusiasm was tremendous in overcoming difficulties and advancing efficiency. The establishment of minewatching posts and reporting arrangements went ahead steadily, the Achilles heel (as ever) being unreliable communications for a Service that had to rely on second-hand equipment, used infrequently.

After two years, with the Service still forming and steadily becoming more proficient and with new sights being developed for greater accuracy, the Admiralty decided it was time that firmer Command and Control arrangements should be made. Accordinly, ACR was instructed to take over responsibility for the Minewatching Service. This time without demur, ACR set about centralising the administration and cutting out many anomalies that had arisen between Areas, but he soon came up against the command problem. For it was the Commanders-in-Chief who trained and employed the minewatching Units and to whom the Units turned when in need. Commanders-in-Chief were quite content for ACR to administer the service so long as they retained ultimate control over 'their' Units. Command Minewatching Officers found that trying to satisfy two different Senior Officers was confusing and very time consuming and it seemed that either the Minewatchers had to integrate with the Reserves or revert to the previous loose arrangement. After two years and much deliberation, in 1956, the Minewatchers reverted to the command of the Commanders-in-Chief with the Commander-in-Chief Home being given a co-ordinating role and the sponsorship of the Service in Admiralty was resumed by the Plans Directorate. It is pertinent that in that same year, which included the Suez Operation and major cuts in the Navy, the Commander-in-Chief Plymouth—against the advice of ACR—set up the first Headquarters Reserve Unit (HQR) at Mount Wise, Devonport. His aim was to augment the small active Service staff at his HQ with trained personnel to support the frequent peacetime requirements for NATO and national exercises. In subsequent years HQR Units were raised separately at all other major Naval headquarters in the United Kingdom.

By 1958, the whole maritime situation was changing. The Soviet Union not only possessed nuclear weapons and their means of delivery but also a Navy which was rapidly expanding into a major deep water fleet. With the growth of the nuclear threat, the perspective of how best to protect United Kingdom ports and commercial shipping once more shifted and minewatching assumed a lesser priority. Minewatchers had shown themselves to be dedicated and resourceful and their coverage had been extended by the introduction of afloat minewatching and the development of better minewatching sights. Now that the threat of a minelaying offensive was receding, and in any event no money was available to upgrade the technology, the Minewatching Service began to be used for other tasks at ports and anchorages. In 1961 the 'Port Party' concept was introduced whereby, in emergency and if not required for minewatching, members would undertake such tasks as manning local craft, car driving, typing and communications.

In the Second World War protection of merchant shipping had taken a very large slice of the Naval effort and the proven expertise of the United Kingdom NCS Organisation had subsequently been recognised and adopted by NATO. Although updated, it had hardly been exercised and most RN and RNR officers with wartime NCS experience had by this time retired. However, dedicated staff within the Trade Directorate had nurtured and trained a core of reserve officers, and complements for NCS and port defence had been constructed from reserve personnel left over after the priority needs of the fleet had been filled. The nuclear threat now imposed a wholly new timescale for transition to war and whilst at this time the naval reserve position was satisfactory, it was becoming evident that personnel required for naval duty at ports should not only be available early and quickly but also that they should be trained and ready for emergencies. These mobilisation matters were under discussion when the Cuban Crisis intervened to highlight their relevance.

So it was that Commander R L Liveing, the Senior Minewatching Officer and a member of the Naval Plans Directorate, suggested that the Royal Naval Minewatching Service should be reformed, using the same constitution, with the primary aim of providing "Port party support for the Local Naval Authority" on a local and immediate basis and retaining responsibility for minewatching should this be required. The Naval Staff adopted the suggestion and recommended that a serving Naval Captain should be appointed as Head of the new Service to provide naval direction and control under the authority of the Commander-in-Chief Home. The title Royal Naval Auxiliary Service was chosen, "auxiliary" denoting the Service's Civilian Volunteer base.

Minewatchers Afloat

Enthusiastic Minewatchers arrive in Poole onboard their Afloat training craft—MFV.95. Minewatchers went afloat to extend the coverage from shore and to watch over the all-important port approaches.

Laughing now . . . . but this group of Minewatchers were often involved in exercises which required concentration, good drill and the stoicism to overcome the discomforts of exposed minewatching sights.

The aim was to plot the splash points of mines from craft anchored in the approach channels. The inaccuracies of their basic equipment had to be accepted—despite obvious limitations.

Her Majesty Queen Elizabeth the Queen Mother attended the first annual Thanksgiving Service for Minewatchers—a tradition established in 1956 and still maintained today.

Chapter Three

Formation, Constitution and Role

An announcement in the House of Commons on 5 November 1962 brought the Royal Naval Auxiliary Service (RNXS) into being. It was not reported by the daily media, only Hansard and two local newspapers recording the event. Perhaps a desire to continue business as usual by the Admiralty and a role for the Service which was not understood by the public, were responsible for the lack of publicity? Indeed, on the surface little changed apart from members' shoulder flashes. At its peak strength the Minewatchers had comprised 121 units of varying size with a total strength of over 4,600; now there were 94 units with a strength of 3,800.

The constitution of the RNXS continued unchanged from that of the RNMWS, and remains virtually the same today: a uniformed, civilian volunteer service administered and financed by the Admiralty. Recruitment, training and day to day running of the Service were delegated to Commanders in Chief and the Flag Officer Scotland and Northern Ireland. Most of the permanent staff of 23 continued in their same Civil Service grades and the posts they held were retitled from 'Minewatching' to 'Naval Auxiliary' as, for example, the Command Naval Auxiliary Officer, the Section Naval Auxiliary Officer and so forth. The Area Training Officers' areas and titles remained unchanged although their responsibilities increased.

Now, as then, before entering the RNXS volunteers are interviewed and on entry undertake to respond to call out on general mobilisation in a full or part-time capacity. Members are entered on List One, the RNXS main stream, in which they are expected to attend and train regularly and to participate in relevant exercises. List Two is comprised of those members who for valid reasons cannot attend regularly for training but whose services it is desired to retain for emergency or until they can rejoin List One. In peacetime, the principles of informality and 'all of one company' adopted for the Minewatchers remain in force. Auxiliaries are not paid for peacetime training or duty but instead are eligible for flat rates of allowances and travel costs to meet out of pocket expenses. The most recently joined naval auxiliaryman and the longest serving volunteer Auxiliary Officer thus receive exactly the same recompense for an evening's training session or participation in an exercise, in recognition of the principle that the time voluntarily surrendered to support the Navy is of equal importance to the members giving it, irrespective of the level or quality of the support given. Advancement or promotion do not bring tangible rewards but rather a sense of achievement for the individual, who must work hard and long to reach the required standards and levels of experience. Badges of grade and qualification bear witness to such achievements. Members are not in peacetime subject to the Naval Discipline Act but, as civilians, conform voluntarily to its principal tenets on a self discipline basis. These principles, which underpin the peacetime ethos of the Service as spare time, voluntary, civilian, unpaid and unarmed, remain as relevant as they were twenty five years ago. They are not, however, necessarily the best basis for the organisation in wartime and this aspect and changes under consideration are discussed in Chapters 6 and 12.

The role of the new Service was to provide port parties in support of the local naval authority and to support the Naval Control of Shipping Organisation in particular with the boarding and berthing of ships at Assembly Anchorages. A much wider knowledge was needed of communications, plotting and NCS procedures and extra equipment such as telephones, radio sets, plotting instruments and vessels were needed for training. There was also the 'shape' of the Service to consider. A number of Units were not well placed for their new responsibilities: some larger ports had several small Units covering minewatching posts whereas now only one Unit was needed to support the Naval Officer in Charge; elsewhere there were no Units at some secondary ports or insufficient Units covering assembly anchorages where minewatching had not been extensive.

The Units' training centres ranged from well found premises in Naval and Admiralty establishments to a room hired once a week in the local pub. The majority of Units trained one night a week at these centres, many of which belonged to the Territorial Army, the Sea Cadets and the Royal Naval Reserve. The infant Service had little of its own and the general lack of permanent accommodation made continuation training difficult.

There was plenty to think about during those early months but, without doubt, the most important factor was the volunteers themselves; to nurture their interest and morale and, secondly, to improve their capabilities to the point where they could temporarily fill the billets of active service personnel in emergencies at the ports of the United Kingdom. The Minewatchers had worked with the sub-

New Service—New Equipment

Technical training is carried out on a model of an inshore minesweeper—these craft provided a major boost for training.

RNXS Members acquired what was for them, new equipment. This presented a major training commitment for teleprinter operators and communicators.

ordinate Home Command organisation as a separate force providing a service. If the RNXS was to succeed, it had to be integrated in the Naval Home Command organisation at Port level and this would require fostering the correct relationships with the Royal Navy for training, administration and supplies. This was where the new-found Head of the Service, a Royal Naval Officer, could play his part. Unfortunately, although approved by the Admiralty, the appointment did not gain financial approval until June 1963. The officer chosen, Commander John Murray, was overseas at the time, but he began briefing for the job in September and took up his appointment as Captain RNXS formally in December 1963.

What sort of people were these members of the newly formed RNXS? They were from a wide cross-section of British society ranging from retired admirals and active professors to clerks and factory workers. Every profession and trade except doctors seemed to be represented. Doctors, certain workers required for key defence jobs and, of course, those already serving in the forces and their reserves were barred from joining. Each Unit had a character and composition of its own, dependent upon the port and its environment and forged by the membership's interaction with local conditions, each other and the resources provided by Admiralty. They shared a great sense of loyalty and service. They had joined out of a desire to help with the defence of the country, a respect and love for the Royal Navy and, for many, a need for comradeship and to be in touch. Many, of course, had served in the Forces in the Second World War; others remembered what had been achieved during that war by corporate and voluntary effort and, underlying all the variety of reasons for belonging to the RNXS, lay the great spirit of voluntary service that is part of the British character and a fundamental strength of the country. The age limits of the Minewatchers had been set at 21 to 65 with extension to 70 in order to draw on a wide field to build up numbers as quickly as possible. The aim had been achieved but it had left the Service with a high proportion of the over-sixties and an overall average age in the mid-fifties.

The "Father" of the RNXS—the late Cdr Bob Liveing RN, who steered the RNXS on its new course.

In the Second World War there had been a huge expansion in the number of small Admiralty craft for servicing the fleet with all kinds of stores and provisions. The vast number of craft with the incongruous title of "The Small Vessels Pool" (SVP) was directed by Admiral Sir Lionel Preston and deserves a history of its own. One of his problems had been in delivering new craft from the small boatyards around the coasts to their operational bases, because of the severe shortage of seamen. Volunteer Admiralty ferry crews were formed, virtually by personal contact and word of mouth, from amateur yachtsmen who could be spared from time to time from their civilian war time jobs. These crews were controlled by a remarkable man, Charles Barber, a tailor in the Fulham Road, London, who was also a wartime Executive Officer in the SVP in the Admiralty. After the war the Admiralty Ferry Crew Association (AFCA) was formed, many of the crews keeping in practice together, should they again be required. Their training was arranged by the Director of Marine Services, the successor of the Director of the SVP, but after the formation of the RNXS it was decided in 1963 to transfer this responsibility. Consequently some 300 active members of the AFCA became List 3 of the RNXS under the leadership of Commander Godfrey Ransome who had served with the SVP during the Second World War. The AFCA's experience of small craft was to prove of considerable value in building up the afloat branches of the RNXS.

With Units scattered throughout the United Kingdom, many hardly known by the Navy and anxious about their future, the sentiments at the launch of the RNXS were well expressed and much appreciated in a 1963 New Year message from the First Sea Lord, Admiral of the Fleet Sir Caspar John:

> "I wish to congratulate you all on a fine achievement—the successful birth of the Royal Naval Auxiliary Service. I particularly wish to thank all members of the old Royal Naval Minewatching Service who over the past ten years have stuck to their task and have built the firm foundation on which this new Service rests.
>
> The Royal Navy Auxiliary Service is a bold, imaginative, and forward-looking concept full of promise and expectation. I know its future can safely be left in your hands.
>
> The Royal Navy is grateful to you for shouldering new burdens in peace and war. I, myself, can pay you no greater compliment than to assure you that the Navy will expect from you the same high sense of devotion and purpose as it does from all other men and women who serve it—each in his own capacity.
>
> Good luck to you in 1963 and always."

CASPAR JOHN

Additional Roles

RNXS personnel join in a Naval Control of Shipping exercise.

The RNXS were called in to support an official visit to Duisberg in lieu of an HM ship. There the author (as Captain RNXS) says farewell to the visiting mayor.

Chapter Four

1962-69 The Years of Transition

XSV Pagham *and seven other inshore minesweepers brought out of reserve provided much improved sea training. A great boost to mobility—and morale.*

On its formation, the Royal Naval Auxiliary Service was viewed by the Admiralty as an important supplementary force for local naval support at commercial ports, where the Admiralty still held the chief co-ordinating role for maritime affairs and where its word carried much influence. The RNXS was as yet untrained but it was available at short notice and on the spot as an organised body to assist in the immediate action required to utilise and protect merchant shipping with the trip wire strategy of the time. Under the direction of naval offices with dual peace/war appointments and reserve officers training in NCS, members would be invaluable in the initial stages whilst mobilisation took place and war complement billets were filled by reserve and recalled personnel. Thereafter the RNXS would carry on a separate and auxiliary service undertaking extraneous tasks and understudying in appropriate billets to release active Service personnel for seagoing duties with the fleet. Thus, whilst prepared to give every encouragement and moral support to their new offspring, the Admiralty did not envisage the allocation of additional funds and instructed that existing votes and supplies were to be used. In those days budget margins were relatively generous and stores and clothing, for example, were available in sufficient quantities to accommodate the needs of the RNXS with no perceptible effect on funding. This was not unsatisfactory for the RNXS as Admiralty Civil Servants were adept and experienced in such ways, enabling the RNXS to be run economically and efficiently at little extra cost.

In 1964 the Admiralty became the Navy Department of the Ministry of Defence (MOD(Navy)) and responsibility for the co-ordination of maritime affairs passed to the Department of Trade. Whilst this had little impact at that time, it tended to obscure MOD(Navy)'s continuing responsibilities for defence of ports and anchorages. The ever growing threat to NATO shipping posed by Soviet submarines and, later, by Soviet surface fleets, caused a major overhaul of plans for the protection of shipping. As a result, training in NCS procedures became the first priority for the RNXS, followed by the requirement to be ready to assist with the evacuation of shipping overseas and with the dispersal of port plant (tugs, lighters and so forth) to safer havens.

Achieving and maintaining the levels of skill required by RNXS members to discharge their new duties in a reliable manner, proved to be the chief problem tackled by the permanent staff for the first seven years of the Service and indeed the problem in some respects persists until the present day. Apart from the Area, later Group Training Officers, whose efforts perforce had to be largely directed to training administration and not instruction per se, there were no full-time professional instructors in the field; Units, unlike their RNR Training Centre counterparts, were obliged to rely on self-training for their one evening per week training sessions. Each Unit selected an 'Instructor' for each branch from within the Unit membership: the member so nominated would in some cases have a service or civilian instructor background but little subject knowledge; in others be knowledgeable in the appropriate subjects but have no experience of giving instruction. Every available instructional resource needed to be tapped, in the event not always successfully, to raise standards in communications. plotting, seamanship, pilotage and engineering. Some of the necessary skills had been carried forward from minewatching days but now the scope needed to be greatly expanded. Inexperienced 'Instructors' were given short courses at command centres or in the RNXS school at HMS *Vernon*. Naval establishments sometimes lent RN Instructors on an ad hoc basis to nearby Units, although too often a lack of continuity largely nullified these well-meant efforts to help. With no proper syllabuses in the early years, with a pool of Instructor talent of so variable a quality and with infrequent opportunities to practice the skills so laboriously acquired, it is little wonder that successful training was a major problem in those early years. Much credit is due to the patience and perseverance of trainers and trained alike in nevertheless steadily advancing the standards and skills of the infant Service.

The raw material to be taught to the members came from diverse sources. For some subjects, manuals were readily available; for others no suitable manuals existed or security classification prevented the issue of those which did exist and it was necessary for the RNXS School to embark on the production of its own handbooks. Syllabuses were simply scaled down versions of those used by the Royal Navy, with the RNXS adopting 'ordinary' and 'advanced' levels as their approximate equivalents for a port party service. Even more of a problem was the teaching of NCS. The manuals were classified and few active service officers had a good knowledge of the procedures. Indeed early in the reactivation of NCS came promulgation of local NCS orders by Commands for their own purposes and instruction, but these too were classified and could not be released to the RNXS. The solution came with the writing of a manual specifically for RNXS Units. Lieutenant Commander Erlend Clouston of the Scottish RNXS staff, a qualified NCS Officer, together with Commander Charles Emerson, the Scottish Command Naval Auxiliary Officer, not only wrote but also illustrated a comprehensive NCS manual which was widely admired and used. They also produced an accompanying handbook of NCS exercises including instructions for their conduct. But problems over security, chiefly with safe custody and promulgation of instructions, were to hamper training for many years and to some extent persist to this day.

The RNXS knew by 1966 what it had to train for and was learning quickly how to go about it. However, it was far from clear what members and grades of personnel were required at ports to support Naval Officers in Charge (NOIC) or Resident Naval Officers (RNO). There had been little revision of Subordinate Naval Shore War Complements since the end of the war although circumstances had changed out of all recognition. Amongst many other anomalies, complements continued to show no less than seventeen Flag Officers in Charge subordinate to the Commanders in Chief, some of them at ports which the RNXS was not required to support! In 1966 the Geraghty Committee, following a review of the Naval Command Organisation, recommended a new peacetime organisation appropriate to the reduced size of the Navy. Implementation of the recommendations was complicated by redundancy problems, by NATO requirements and by the need for co-ordination of the three Services. The 1966 naval manpower strength plus all reserves amounted to a quarter of the naval strength of August 1945, so much streamlining of the outdated complements was inevitable and long overdue. Meanwhile the RNXS staff, guided by the Naval Plans Directorate, unilaterally decided manning scales for Units and began the process of adjusting the size and location of Units for the type of naval support envisaged. It was not until July 1969 when the Naval Home Command, with Area Flag Officers, was established, that the Commander-in-Chief Naval Home Command was directed by MOD to revise the local naval command arrangements based on the reduced manpower available, allowing for only twenty five NOICs and seventeen RNOs and incorporating the United Kingdom NCS Organisation. It was expected that a definitive strength of RNXS members would thereafter be soon available but the RNXS staff were to learn that plans, involving different authorities and necessarily taking account of economic and industrial changes of great significance, were imprecise and ever changing. So it became necessary for the RNXS to take the long view, if training and administration were to be managed economically and continuity maintained. Units could not be closed, nor opened up, nor disbanded at short notice. Yet where there were important changes which directly affected the RNXS, care had to be taken to explain them to members and for the Service to respond to the needs of the Navy. For example, 'flexible response' replaced the NATO wire strategy in 1968 and a very early requirement arose for NCS with some members needed well before mobilisation.

The growth and activity of the Soviet fleets, naval, mercantile and fishing, constituted a growing threat not only to shipping but also to ports and their approaches through the possibility of clandestine activity. These factors led to the need for much better preparedness, for which regular exercising was a fundamental ingredient. The RNXS took part in National and NATO exercises, initially on an opportunity basis; then, as confidence grew, exercise schedules incorporated RNXS members' participation from the outset. Such exercises were also of great importance in that they enabled NCS plans and procedures to be tested operationally. RNXS exercise support was popular with Naval authorities as it reduced the need for active service personnel to participate, thus cutting down on drafting turbulence and demands on uncosted manpower.

It was not however all sweetness and light during those early years. Too often, little warning of exercise requirements was given showing little appreciation of the difficulties auxiliarymen had in getting time off from work or in making domestic arrangements for the duration of the exercise. Many officers and senior ratings were uncertain how to deal with auxiliarymen, their approach ranging from the avuncular or head patting to the very blunt or almost rude. Members expected to be taken seriously and wanted to be treated in the same manner as naval personnel but with allowance being made for their different backgrounds and ages. There was a lack of expertise in certain cases and the RNXS was sometimes at fault in involving members in exercises before they were ready. Indeed the difficulties of providing adequate instruction within Units led to the use of exercises for instruction which was perfectly satisfactory while the exercise aims were limited and the pace could be controlled to suit instruction. Later it led to confusion in some key exercises where the twin aims of testing operational plans and basic instruction were incompatible. Conclusions on operational matters drawn from such exercises could prove unsound: for example, it would be concluded that procedures were wrong when it was truly training at fault or vice-versa.

Live exercises were always popular, while for so-called paper exercises, popularity varied with the degree of activity injected. Quite often large scale paper exercises lacked adequate preparation and this led to long quiet periods. Members became bored and wondered why they were closed up. Local 'drill' practices were prepared in a standard form nicknamed "Bartholomew" to fill these quiet periods, but proved only partly successful through lack of directing staff experience in organising practices at short notice.

Afloat training was assisted greatly by the allocation of the first four Inshore Minesweepers in 1964 to bolster the MFVs inherited from the Minewatchers. Despite gloomy prognostications that auxiliarymen would have difficulty in maintaining these vessels to provide adequate training, with the help of the Port Auxiliary Service and three Engine Room Artificers and Mechanicians, these Auxiliary Service Vessels (XSVs) ran regular training programmes from March to October each year from their base ports around the United Kingdom. Although for some years there were serious shortages of afloat personnel, these vessels were always in demand. Fortunately there were usually just enough experienced crews for exercises, guard duties and cruises; few requests were refused.

Thus, as RNXS members slowly gained experience, mutual respect and understanding grew between them and active service and reserve personnel. Gradually, too, an esprit de corps emerged with their NCS officers, enhanced by the excellent morale at RNXS Units. The good morale was supported by a caring although not always adroit administration, and advanced through such maritime experiences as the clearance of oil following the "Torrey Canyon" accident and "Proud Warrior", the first major national exercise to attempt a comprehensive test of the United Kingdom subordinate Naval Home Command Organisation for many years.

On 27 July 1965, Her Majesty the Queen approved the institution of the Royal Naval Auxiliary Service Medal, to be awarded on recommendation after twelve years continuous and meritorious service. Morale was greatly boosted by this gracious recognition of the considerable dedication and domestic sacrifice made by RNXS members with no tangible reward. Approval had taken four years to achieve, although not due to reluctance or opposition. The proposal had begun its passage through the offices of State as a Service Medal for Minewatchers; had then been laid on one side whilst the transformation to the RNXS took place; was forgotten for a year due to pressure of events and finally was brought to fruition in a burst of consultation and decision-making over which type of fouled anchor and which colour ribbon were appropriate or historically correct. The first awards of the medal were made in 1966 and, as Minewatching service was allowed to count, the lists for the medal in that year were long and the medals office had difficulty in keeping abreast of the engraving of the recipients' names on the rims of their medals.

In February 1968 Commander Bob Liveing, the Senior Administrator and midwife for the birth of RNXS, died suddenly. He had achieved a great deal but he had not been able to see the RNXS as firmly established as he would have wished. His knowledge and experience, particularly of the MOD, were sorely missed and the then Captain RNXS, Captain David Johnston, had a very hard six months coping with his own and the Senior Command Naval Auxiliary Officer's jobs in Portsmouth and London until a relief was appointed. The first Captain RNXS, John Murray, serving at that time on the staff of the Admiral Commanding Reserves, was prevailed upon to retire early from the Navy to take up the vacant post and was eventually appointed in September 1968. The RNXS being a unique organisation, previous knowledge was important and Murray was also fortunate in that few of the Civil Servants concerned had changed so he was able to gather up the reins in time to make financial provision for the following year and embark on the process of shepherding the RNXS through the next sixteen years.

All Part of the Local Scene . . .

PNXO Humber receives the crest of the City of Hull to cement relationships between the Hull Unit and the city.

Money raised by RNXS Units in South Wales is presented to the local branch of the RNLI.

Chapter Five

1970-77—The Years of Advancement

XSV Loyal Moderator—*one of the eight Fleet Tenders specially adapted for RNXS training.*

1970 marked a further stage in the development of the young Service. In September 1969 a large scale national exercise nicknamed "Proud Warrior" in which nearly seven hundred members participated, had been held to test the revised local Home Command organisation at ports and anchorages. Although only 35 (out of the 42) NOIC and RNO were closed up, the attachment of NCS staffs to the NOICs put a heavy burden on inadequate communication facilities. It was only by the use of unorthodox and potentially insecure methods, and by local initiatives, including the use of despatch riders between ports, that some semblance of order was maintained. In this, the first exercise conducted by Commander in Chief, Naval Home Command with the new organisation of Area Flag Officers, some fundamental lessons were re-learned. In short, reactivating an organisation, itself much altered, after some twenty years in limbo for policy reasons is far from simple and can be expensive. Few of those involved appreciated that it would take a further eight years or so to reach satisfactory standards. The specially appointed staff officer for the exercise was Commander David Farquharson, who, on completion, was appointed Captain RNXS.

From Proud Warrior came recommendations for an overhaul and modernisation of communications, better NCS procedures, improved training and the setting up of permanent local headquarters. Home defence was a low priority at the time and it was difficult to find money for improvements to the infrastructure. Funds were, however, available for training, the principle being that the training of reserves and auxiliaries, although taking time to come to fruition, was essential for readiness but that most infrastructure needs could be met in emergency by compulsory purchase or requisition. The RNXS training centres, although sparsely equipped, were beginning to improve overall, often through the efforts of the members themselves. Not only did they provide equipment and furniture at their own expense but also they gave their spare time to "DIY" long before it became a generally accepted way of domestic life. The money for RNXS premises was prised from Minor Works Grants and contingent funds; competition with other desirable items on Flag Officers' lists meant that only about four new training centres per year could be found reconditioned or updated. It was against this background that in April 1970 the Commander-in-Chief, directed by MOD(N) to propose a revised United Kingdom subordinate Naval Home Organisation, declared his intention to make use of training centres as local headquarters whenever practicable.

This decision proved to be a turning point in the arduous journey towards satisfactory accommodation for RNXS Units. Flag and Senior Officers now had a dual interest in training centres which doubled as local headquarters and the case for better facilities, whilst it never became easy, could be argued with greater fluency. Furthermore, the RNXS became the custodian for another 'asset' to add to the provision of trained manpower for NCS support ashore and afloat. Interestingly, a debate arose at this time at CinC's Fort Southwick Headquarters over the naming of local headquarters. The title Port Headquarters had been used for some years, largely for convenience in exercises, but the Commander-in-Chief himself suggested the title could be misleading because in many commercial ports the Harbour Offices were termed Port Headquarters: only in naval controlled ports was the title appropriate; elsewhere the title "Naval Officer in Charge Office", although clumsy, was accurate. However, since at that time NOICs were predominantly concerned with NCS matters in commercial ports, the title 'Port Headquarters' carried the necessary air of authority, was recognised by most of those involved and it was important to enhance the visibility of local naval organisations, for want of a more precise term the title was retained.

The seagoing element had made steady progress during the 1960s with the adapted inshore minesweepers (IMS) taking part in the regular annual exercise programme. Auxiliary service vessels (XSV) were used for boarding, delivering despatches and stores, and occasionally berthing duties for convoy assembly and departure. When there was a lack of merchant vessels to make up a convoy for minor exercises, XSVs sometimes acted as convoy ships to give live training to the shore headquarters. It was also useful experience though hard work for the crews, who were often required to change their role several times during an exercise. XSVs were entrusted with other tasks apart from exercises; attending fleet reviews where they undertook patrol and despatch duties, guard duty at yacht regattas and various local tasks in support of the Navy. While efficiency and capability were improving, the numbers of afloat personnel were declining in 1969/70 due partly to a retirement age bulge but also to the difficulty in providing sufficient afloat training opportunities. With twelve craft, eight ex-IMS and four MFVs, and an annual training season from March to November, in theory there were sufficient billets to maintain a strength of 800 and allow for wastage. In practice, availability of the twelve craft was only moderate due to the low dockyard priority afforded to them. Refits and dockings were always planned for the winter months but often, due to higher priority work in HM ships, small craft programmes fell behind. The MFVs, although much loved by many and in particular their coal fired galleys by the cooks (!), were slow (maximum speed 6½ knots), old and unreliable. It was therefore a mixture of pleasure, pride and relief that greeted the approval of a programme for the replacement of all twelve craft by new Loyal class fleet tenders in 1971. These type 'C' fleet tenders were specially adapted for RNXS use with good facilities, robust and seaworthy. However, they had no gyro compass, the derrick arrangements were unseamanlike and access to the main accommodation was exposed in bad weather. The first two to enter service were Loyal Governor, allocated to the Plymouth area in 1971, and Loyal Factor, to Medway in 1972. Within a year both had been deployed to Northern Ireland under the White Ensign for anti-terrorist inshore patrols. In 1973-74 the Loyal Moderator, Loyal Chancellor and Loyal Proctor were accepted and soon proved their worth, despite the limitations imposed on the two latter craft through the short-sighted economy of equipping each with just one large generator instead of two moderately sized ones. The arrival of the Loyal class gave a fillip to engineering confidence and as with the IMS, it was again learned that the more they were used, the better their reliability. Indeed statistics showed that the most likely period for breakdowns was following refits and maintenance periods.

By 1972 the ferrying commitment had been 'written out' of current plans and the Admiralty Ferry Crew Association (List 3 RNXS) was, to widespread regret, disbanded. Despite the AFCA skippers' valuable assistance in training the RNXS as a whole, as well as their own association members; retention of a separate list was no longer justifiable and it was becoming increasingly difficult to accommodate the differing requirements of List 1 and List 3 training. The problem was accentuated because almost all of the 300 AFCA members were concentrated in the SE Area and the training load could not be spread. Those eligible to do so were invited to transfer to Lists 1 or 2; in the event 98 transferred, most of the remainder living too far away to be able to undertake regular training.

While shore training continued unabated throughout the period and larger numbers qualified, exercises indicated that standards were not rising enough to meet an increasing requirement and that training, except for exercises, was often dull and uninspiring. A staff re-organisation was undertaken in 1972/3 to increase the number of training officers from 5 to 12, with a compensating reduction in the number of administrative officers from 11 to 4. The drastic reduction of administrative effort was only made possible by the recently introduced centralisation of the control and distribution of MOD(N) stores and clothing, encompassing RNXS needs within the overall requirements of the Navy. Additional administrative burdens were placed on Heads of Units and accepted on the grounds that they gave useful training; the subsequent growth in administrative requirements could not be foreseen at the time but has in recent years reached the limits of toleration for a 'spare time' Service. This reorganisation took two years to implement but the improvement in training standards was profound and well worth the effort required despite the administrative consequences. With an average of seven Units per training officer instead of seventeen, supervision, control and leadership were greatly improved, interest increased and wastage reduced. A further small but significant step was taken by arranging to send RNXS volunteer instructors to naval instructional technique courses. The courses of three days, later incorporated in senior courses at the RNXS school, were highly beneficial in improving the quality and control of instruction.

In 1974 MOD(Navy) initiated a review of the reserves to assess whether their structure, numbers and readiness met the requirements of the Navy. The Headquarters Reserve had recently been incorporated into the Royal Naval Reserve and the RNXS, being a somewhat similar organisation also comprised only of volunteers, was included in the Review. Under the chairmanship of Rear Admiral Geoffrey Mitchell, the Review Committee consisted of a Senior MOD Civil Servant, a Captain Royal Navy and two Commodores RNR. It sat for five months, travelled to many divisions and Units and heard a large amount of evidence. A long and informative report with many recommendations was produced which coincided, regrettably, with the announcement of large Defence cuts. Consequently, while not discredited, the conclusions and recommendations were somewhat muted and those involving extra expenditure had to be modified. In fact, although the report found little to criticise, the RNXS could nevertheless not escape the 20% cut in manpower and craft imposed across the board on the Navy. The reduction of 500 in manpower strength was achieved within two years by limiting recruitment and through normal wastage. No Unit had to close as a direct result but

An Era of New Professionalism

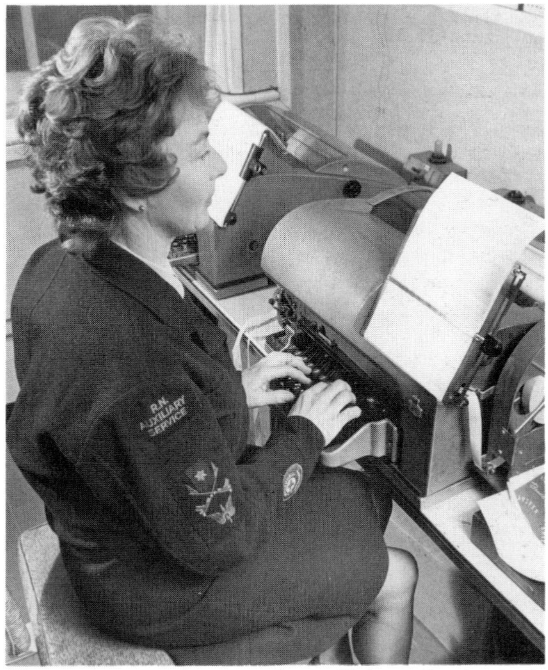

New equipment and new people provided new impetus to the Service in the 1970s.

Rear Admiral Mitchell (centre) visited RNXS Units throughout the country whilst conducting his 1974 review of the RNR and RNXS. Money was short, but the RNXS survived in a healthier state as a result.

it was a disappointment to have to reduce numbers when it was uncertain how many might be required in emergency and because complementing for war billets still awaited definitive action. The reduction by two to an allocation of ten training vessels was serious when the seagoing component was already under strength. In the event, the cuts acted as a spur and, through improved organisation among afloat sections and tighter allocations of craft throughout the year, both standards and numbers of afloat personnel improved. The allocation of five more Loyal class tenders in lieu of IMS between 1976 and 1978 also greatly enhanced availability.

Unit training was much improved and strengthened in 1976 with the introduction of 'Objective Training'. Objective Training was developed and introduced in Naval Training Establishments from the early seventies and proved to be an effective method of analysing what needed to be taught at what stage, by whom and to whom, thus eliminating waste of training effort and resources. A full assessment of the required operational performance of each grade and qualification is broken down to separate identifiable units of skill and knowledge. These separate units are then converted into a syllabus and the syllabus in turn chopped up into lessons, with objectives and standards to be achieved for each lesson. Instructors' notes are written to assist the Instructor to prepare the content of his lessons.

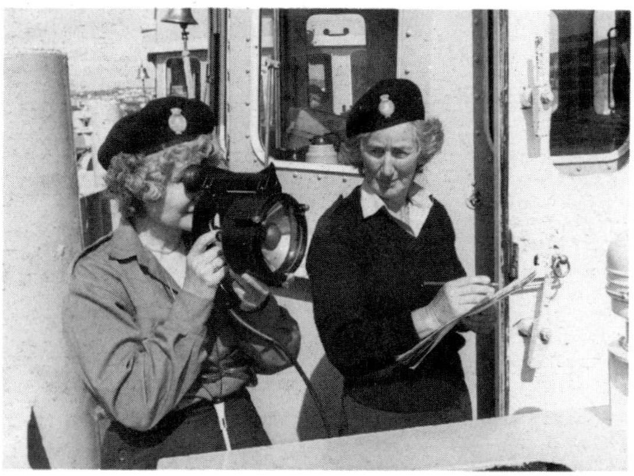

Ladies make up about 25% of the RNXS personnel—and serve both ashore and afloat.

Not so pretty but all good lads . . . . !

Captain Gerry Andrewes, the newly joined Captain RNXS, had been closely involved in the development of Objective Training during his previous appointment in CINCNAVHOME's Training Division and he saw that it would be ideally suited to the RNXS. Widely scattered Units, out on their own with willing but inexpert Instructors, few facilities, variable training documentation and all too rare practical guidance from the permanent staff, needed all the assistance they could be given and Objective Training documentation gave far more than any previous self-help training method. Its introduction involved an immense amount of work at the RNXS school and the documentation took two years to produce and promulgate while at the same time routine qualifying courses continued. The Service remains indebted to Commander David Farquharson, Lieutenant Commander Bill Harding and his successor Lieutenant Commander David Jeans for their efforts in this task together with the naval and civilian staff at HMS Vernon who supported them. It was a milestone in that thereafter RNXS members could be expected to be trained and qualified to approved standards of skills, knowledge and experience.

A further advance achieved by Captain Andrewes was in formulating Training Objectives and Requirements for Units (TORFU) and getting them accepted by Units. The germ of the idea had started in the Plymouth Area by Commander Metherel for recruiting but a TORFU goes much further, analysing a scheme of complement with other area and local commitments to establish a total trained requirement (TTR) for each Unit. The next step is to compare what personnel each Unit has (the current bearing) to provide feasible annual recruiting and training targets, taking into account resources, qualifications and numbers. The advantage lies in providing a realistic target to aim for in the next 12 months. Previously, time and effort were frequently wasted in recruiting and training for estimated requirements or even convenience. A TORFU, while accepting that the Total Trained Requirement might never be reached, is regularly adjusted in the light of new developments or experience, and provides an authoritative reference and guide for the Head of Unit.

As the efficiency and capabilities of the Naval Control of Shipping Organisation improved through the first half of the seventies, so the demands for more reserve manpower for port headquarters and offices at ports and anchorages grew to an extent that concerned MOD for financial reasons. A working party was set up in the summer of 1976 under Captain Peter Hicks-Beach, the retiring Captain RNXS, to develop a revised and economical NCS

Improved training and standards enabled the RNXS to be seen in foreign waters. Here, two smart Loyals visit Bergen on a training cruise.

Organisation. The working party had a difficult task in a changing maritime environment and had to undertake some twenty years of updating which should have been done year by year. The working party recommendations included the need for regular and flexible updating and the adoption of detailed criteria for complementing which fully considered the RNXS role. Whilst the protection of shipping within ports and anchorages was given adequate consideration, the terms of reference of the working party did not unfortunately include the defence of ports and anchorages (DEFPA) and although Captain Hicks-Beach attempted to rectify this ommission after completing his report, the subject was under consideration in MOD at policy level and he was not able to make any progress before he retired. Although it may be argued that financial stringency and differing priorities necessitated the twin tasks of NCS and DEFPA being separated at that time, it is difficult now to see any logic in doing so.

By 1977, through the performance and dedication of its members and much sensibly directed hard work by its staff, the RNXS had grown from a 'come in handy' force to a trained and organised volunteer service ready to fill war complement billets at ports and anchorages. Even in the Navy it remained little known and seldom featured on the national scene; activity focussed at Unit level and recruiting was conducted on a local basis. When involved on a national scale, as in major exercises or newsworthy events, the Service might get mentioned or be given credit but few outside the NCS fraternity knew its worth or significance. This did not bother the majority of members, who preferred instead to be appreciated without fuss by their own Flag and Senior Officers. The more percipient members and staff were aware that the dangers of a setback, resulting from over-confidence or taking on more than could be handled, were very real for the youthful RNXS.

Chapter Six

1977-1984—The Years of Consolidation

The year of the Queen's Silver Jubilee in 1977 marked a further stage in the development of the RNXS. The event was celebrated by Members at their Units joining local Organisations and Services in parades and shows of every description. The standard of turnout was high and in their improved battledress style of uniforms RNXS Members were undoubtedly a source of local pride. The most notable event of the year was the Fleet Review at Spithead, for which all ten RNXS vessels foregathered at Portsmouth from their scattered bases and were organised as a flotilla under Lieutenant Commander Bill Harding. The seven IMS and three fleet tenders took part from assembly to dispersal, a period of five days during which they undertook a variety of tasks, from delivering mail and provisions and carrying numerous cadets and school parties on tours of the fleet, to policing and patrolling on the big day when Her Majesty in HMY *Britannia* reviewed the fleet. The large number of small craft sightseeing made the latter tasks difficult but the RNXS crews acquitted themselves well with skill and patience. Throughout that week the vessels were run and managed with a degree of professionalism that delighted everyone involved; there were no breakdowns of any significance and no damage to the RNXS vessels.

The only untoward incident occurred on the day of the Review, when an IMS manoeuvring astern of HMY *Britannia* at anchor, in endeavouring to keep small craft clear, was carried down by the tidal stream across the bows of the Trinity House Yacht *Patricia*, also at anchor. The *Patricia* was damaged above the waterline but the IMS suffered no damage. After the Review, the vessels from northern and western bases took the opportunity to make visits to the Continent before they returned home, having been away for up to four weeks. They were soon back to their customary task of advancement training and in the autumn most were engaged in the annual NATO exercise.

By 1977 the RNXS ability to support the Navy locally in Home Defence ashore and afloat was not disputed but somehow, in the Queen's Silver Jubilee Year, seeing the RNXS perform on the national stage gave much

Training—The Rough

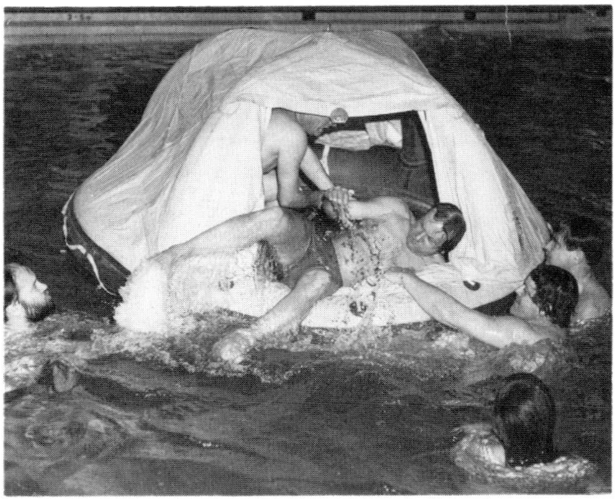

Liferaft drill is carried out by all RNXS personnel who serve afloat.

Even the ladies have to undergo Damage Control training.

With the Smooth

RNXS ladies from Bristol and South Wales.

confidence and a realisation that the Service had matured. For the next two years the RNXS was content to consolidate this progress; nevertheless, prospects for greater responsibility for Officers and senior auxiliarymen began to be discussed. The first steps towards making places on NCS Courses open to RNXS Officers were taken and, as the involvement of RNXS personnel grew, concern arose over their status and rights under emergency legislation. So many factors had changed since the creation of the Service; perceptions of peace and war, and the boundaries between them had blurred as the cold war intensified and terrorism posed increasing threats to internal security. The RNXS as a civilian uniformed force was dependent on mobilisation orders for activation, although in an emergency the Secretary of State for Defence was empowered to call on volunteers selectively before mobilisation. While this met operational needs, it did not protect the volunteers' jobs, security, pay, pensions and other benefits which they might justifiably have expected. Several attempts were made during the seventies to include the RNXS in legislation updating the rights of Reserve Forces but these met with a certain amount of opposition and in the event the proposed legislation failed to progress into law through lack of Parliamentary time. The majority of volunteers were not troubled by these uncertainties, those who gave them any thought believing that legislation might make their position more difficult and that their aim should remain to deter war and aggression by being ready and capable of responding rapidly in an emergency or against a threat. The lack of progress over legislation made improvements to the status of volunteers slow and increases to their out of pocket allowances were hard to achieve at times of high inflation and financial stringency. Nevertheless, it should have been possible to acknowledge the skills and experience now offered by volunteers by at least maintaining their meagre allowances in line with inflation but priority for home defence was low and there was reluctance to seek disproportionate increases for the Auxiliaries compared to the Reserves because, with the tighter budgetting then being enforced, this could have led to either cuts in manpower strength or reductions in equipment. Somehow too, the RNXS permanent staff had slipped a grade in the civil service re-organisation of the 1960s and 1970s and it took until 1980, a three-year struggle of reviews and inspections, to recover the original grading.

At the unit level, these administrative and policy concerns appeared remote and happily there was plenty of activity and interest with training and exercises to occupy the volunteers. Afloat, the re-equipment with the Loyal Class Fleet Tenders was completed by the delivery in 1978-79 of five modified (Batch III) vessels. Having internal access to the forward mess and living accommodation, these vessels featured greatly improved comfort and safety through eliminating the need to use exposed hatches in bad weather. They were also built with improved steel plating to reduce the rusting experienced with the earlier batches. Two IMS, the Portisham and Shipham, were retained to maintain the allowed total of ten craft, in hopeful anticipation of the return of the Loyal Governor and Loyal Factor (temporarily renamed HMS *Alert* and *Vigilant*) from active service in Northern Ireland. To gain sea experience, for navigational training and for practice in operating their vessels over longer periods without support, out of area cruises were encouraged and XSVs undertook voyages across the North Sea to Norway, Denmark and Holland. With training and exercise requirements as the first priority, occasional support for HM Ship visits to seaside towns and duty as guardships for yacht races, regattas and other local events, the annual programmes of the XSVs were soon replete. A marked improvement in serviceability, making more seatime available, was one of the benefits of bringing the new Fleet Tenders into service.

The use of the syncro-lift at Rosyth reduced maintenance time and increased the availability of craft for training.

An interesting assignment fell to the Loyal Helper in July 1980. She was ordered to Ostend to support three HM Ships—*Norfolk, Juno* and *Bildeston*—anchored off the port and taking part in the Naval Review to celebrate Belgium's 150th Anniversary. The Loyal Helper was employed landing libertymen and delivering stores and provisions. Unfortunately the three days were spoilt by gale force winds; indeed for much of the day of the Review small craft had to remain in harbour. Nevertheless, despite the hazards of berthing alongside the ships and of embarking and disembarking libertymen, she carried through her mission with only minor weather damage to HM Ships' paintwork and her rubbing strakes and was rewarded by the Admiral with a huge iced cherry cake delivered by helicopter!

A unique opportunity had occurred earlier in 1980 to learn at first hand the working of the Royal Navy. The Captain of HMS *Hermes*, Captain John MacKenzie, offered to take RNXS volunteers in his ship for a cruise to the West Indies from January 15th to March 25th. In spite of the short notice some 20 volunteered and fourteen actually went. They were allocated their own mess (the Sergeants Royal Marines) to run and clean under the guidance of the divisional officer, Lieutenant D. Todd. They rotated around the departments for instruction and assisted with the daily running of the ship. Places visited included Trinidad, Curacao, New Orleans, Pensacola, Fort Lauderdale and Bermuda plus several small islands in between ports for picnics. The volunteers were able to get ashore to see the sights but, whilst enjoying every minute of the cruise, found there was also a serious side involving considerable effort in exercising as well as maintaining a large ship. The volunteers gave an excellent account of themselves, receiving many thanks for their help and winning the Captain's prize, a cake, for the best kept mess.

While from 1976-79 the standards of performance of auxiliarymen had shown probably the fastest growth rate ever under the firm guidance of Captain Gerry Andrewes, the Captain RNXS, it has to be admitted that the RNXS remained little known, even in the Royal Navy. So when Captain Peter Kimm took over he made it one of his objectives to get the RNXS better known and understood. By many visits throughout the country and through the use of his eloquent pen, he succeeded in pushing the Service more into the limelight. The advantages were soon apparent in such diverse benefits as improvements to uniforms and trips to the West Indies but there were also added responsibilities, one such involving what came to be known as the Yachtsmen's Reserve Saga. The Naval Staff for some years had pondered on whether and if so, how, private yachts and motor boats should be organised in an emergency to support the local defence of ports and anchorages. The phenomenal increase of small yachts during the 1960s and 1970s had been matched by a decline in the UK fishing fleet. The RNXS staff was invited to propose a scheme based on its existing organisation but it soon became apparent that to impose yet another volunteer organisation onto existing volunteer units would not work satisfactorily. The recruitment of yachtsmen (including women) and the selection of suitable boats for the tasks of surveillance and general communication duties would require additional experienced staff. Something could be achieved on a local and decentralised basis but there would be a not inconsiderable cost in administration and annual training to provide a minimum of cohesion and readiness. A comprehensive reserve of yachtsmen on a national scale was not a practical proposition: the costs involved would not have been justified by the advantages gained. Local boatmen and yachtsmen in their small craft did magnificently at Dunkirk in 1940; would they have been able to achieve more had they been organised previously, instead of by radio on the day? Perhaps, but no one can forecast future emergencies and if such reserves are not properly supported and exercised, disillusionment soon sets in and is contagious. Without considerable additional staff, it was beyond the resources of the RNXS to cope even on a local basis, so the proposal, having been explored, was put on the shelf, whence it still emerges from time to time to be dusted off and returned. Meanwhile the RNXS was resolving its own problems.

Ashore the struggle to achieve a high degree of efficiency in Port Headquarters communications continued, always difficult for volunteers because efficiency depends on constant practice. Accordingly the aim was set at providing capable personnel, up-to-date with procedures and requiring only a 24 hour 'warm up' to operate satisfactorily. Teleprinter and radio sets were in limited supply however, and training programmes required careful planning and dovetailing to make maximum use of such equipment as was available. Success in resolving these factors, together with the co-ordination and instruction, briefing and sympathetic support provided by the Navy and RNR, enabled RNXS communicators to achieve very satisfactory performance standards. In the seventies there had so often been communication problems at the start of an exercise through errors and lack of practice, although by the end of an exercise it was encouraging to see it all working satisfactorily. By the end of the decade many of the troubles had been overcome and it was heartening to watch RNXS communicators closing up at their Port Headquarters to begin operating with a confidence borne of ability.

The Operations Branch, formed through the amalgamation of the Plotting and General Duties branches, had achieved a satisfactory standard in the basic skills of displaying and filing information on ships required for NCS and local defence purposes, but progress towards high standards of efficiency was delayed through frequent changes in procedures and the lack of a uniform method of information display in Port Headquarters. Changes in procedures were necessary to reflect the refinement of NCS doctrine but security requirements hampered the promulgation and teaching of new procedures, leading frequently to confusion. Although it might be known that a new procedure was to be introduced, exercises were undertaken regularly using the superseded version.

While no two ports could have exactly the same

operational commitments and certainly no two Port Headquarters were alike, nevertheless uniformity in methods of information display was greatly to be desired since those using the information could be from ships, external posts or other headquarters. Moreover, a common method saves time, reduces error and is a blessing for teaching. This lesson was only slowly learned and regrettably, during much of the seventies, too many pet schemes were allowed to operate for too long. When, eventually, a common method was selected, the Port Headquarters handbook was written and the necessary display boards and equipment provided. There was an upsurge in the efficiency of the operations branch which led to senior auxiliarymen undertaking more responsibilities. Also, by 1980, electronic methods of display and data processing for all Port Headquarters were already in prospect and some pioneer equipment was on trial.

To give witness to the improved status and commitment of the RNXS, it became possible in 1979 to begin a major change in uniform. In the continuing rationalisation of naval uniforms, the blue serge working dress worn by the RNXS was to disappear, having been replaced in the Royal Navy for many years, latterly by the heavy wool jersey working rig. On its own, this rig was inadequate for the RNXS role, so the approval to adopt the naval uniform jacket style of dress, distinguished by the appropriate RNXS badges of rank or qualification, was greeted with much pleasure. Berets were retained (to the dismay of many) but the cost of peaked caps and WRNS hats was at that time deemed too high. Thus arose a major departure in style, because hitherto all auxiliarymen had worn the same working dress, although ladies wore skirts of the same material. Thereafter, auxiliary officers wore Naval or WRNS officer uniform, auxiliarymen wore Naval class I uniform or Wren's uniform with the only common item for all was the beret.

The reverberations from the 1981 Defence White Paper ran through the Navy as far as the reserves and the auxiliaries. After their recent progress it seemed at first that the RNXS were in for a period of retrenchment due to financial stringency but events were to prove the reverse. There were two main reasons; the higher priority being given to Home Defence and the declining strength and availability of the regular naval reserves. With the nuclear stalemate, there had appeared little need to defend our ports until the ubiquity of the Eastern Block maritime forces was seen to pose a threat to them. The capability of those forces to disrupt maritime traffic by various means became apparent and the use of UK ports for reinforcement and for responding to threats in time of tension was essential. Plans for the Defence of Ports and Anchorages (DEFPA) were long overdue for revision. The reductions in active service manpower demanded by the Defence White Paper led to another review of naval reserves to determine whether there would be sufficient for future new and revised commitments and whether better use of reserves and auxiliaries might be made for peacetime support. There were no concise answers to these questions but, while the current overall reserve state was healthy, shortages were arising in some categories which the RNXS with recruitment and training might fill. It was foreseen that increasing the role of the RNXS would involve yet higher standards of performance and training leading to more centralised control. An improved supply of modern equipment would also be necessary to achieve the standards and gone forever would be the "come in handy" syndrome for the Service.

Geminis replaced the 10-foot dinghies in the training craft. New equipment meant new experiences and greater efficiency.

The permanent staff had been restructured successfully in 1973 but the part-time staff officers' structure had continued virtually unchanged since the minewatching days. It was recognised that to provide an increased degree of professionalism the permanent staff had to be increased, with more officers based at key positions for NCS and DEFPA. These tasks were being undertaken by part-time officers, in theory borne for administrative supervision and with conditions of service that allowed only an average of 10½ hours paid service a week. Most in fact did much more and their hours were often used up completely by participation in exercises. To alter their conditions of service to provide longer hours, if they could accept them, meant a complete change of category from "temporary" to "semi permanent" part-time. This was considered to be an impracticable method of achieving the aim. The majority had given loyal and valuable service over many years, some as much as twenty two years; many would not welcome such a change and some would be ineligible. The matter was still under discussion early in 1982, at a high and confidential level, when suddenly the

Falklands War intervened and all ideas of restructuring were put back on the shelf.

The Falklands campaign was a quiet and anxious period for the RNXS: offers of help were made but, apart from minor communication support, there was little a service confined to the United Kingdom could do to help. The consequent cancellation of exercises provided a spur to basic training ashore and, afloat, all RNXS craft were as active as possible ready to undertake suitable tasks to fill gaps. Like all previous wars, the Falkland campaign had a remarkable effect in speeding up the administrative processes. It was really heartening for those working on the sidelines to witness the speed and efficiency with which maritime business was transacted and best of all, from the point of view of the RNXS, some Home Defence aspects were carried forward on this tide. Plans for DEFPA, which had been under scrutiny or review for up to five years, were brought forward, calling for increased manpower at a time when regular numbers were reducing and therefore seeking an increased contribution of manpower from the RNXS. The plans included detailed requirements for requisitioning craft in an emergency for patrolling at ports and boarding merchant ships and for faster, more modern training craft. Port headquarters communications were to be modernised and data processing introduced. It was evident that an RNXS staff restructuring was essential if the Service was to react to these plans in good time. There could be no standing still, it was time to move forward again, despite all the problems that needed to be solved.

1983 saw the twenty-first anniversary of the formation of the RNXS and many local celebrations were held by Units throughout the country. Official celebrations culminated in a staff party held in the wardroom of HMS Vernon at Portsmouth, the alma mater of RNXS training, on 4 November during the Annual Staff Meeting. Admiral Sir Desmond Cassidi, Commander-in-Chief, and Lady Cassidi, were the guests of honour and amongst other guests, all but one of the previous Captains RNXS were able to attend. Lady Cassidi cut the magnificent birthday cake and the First Sea Lord sent a congratulatory signal. The RNXS had indeed come of age.

At the same meeting, Staff Officers were informed of the plans to restructure the 'Field Staff' of the RNXS, the

The Coming of Age Party . . . . at HMS Vernon. Lady Cassidi cuts the all-important cake!

RNXS Units often participate in local parades, exhibitions and shows.

Captain Peter Kimm—Captain RNXS from 1979-82.

results of the deliberations begun before Operation Corporate intervened, then resumed and given added impetus by the implications of the 1981 Defence Review and the upgrading of Home Defence. In all but four (later two) port groups of the United Kingdom, a new style of officer, the Group Naval Auxiliary Officer (GNXO), with both training and administrative responsibilities for the Units within his Group, was to replace the current Group training Officer, with training responsibility only, and the part-time Port Naval Auxiliary Officer (PNXO) with administrative responsibilities only. Four part-time PNXOs would remain and there were proposals (later implemented) that volunteer (unpaid) PNXOs might be introduced to assist the GNXO, particularly with PHQ duties, and to provide a knowledgeable and experienced local input. Restructuring was to take four years so that the transition could be achieved in as smooth and painless a manner as possible. There had been much heart-searching over, particularly, the abolition of the majority of the part-time PNXO posts: the long and invaluable services of those that had held them had underpinned the efforts of the permanent staff in building and developing a viable Service and those PNXOs who chose not to remain in a restructured staff, or for whom there was no appropriate task, would be greatly missed.

In 1984 when Commander John Murray retired after sixteen years as the Command Naval Auxiliary Officer, he left his successor (Captain Mike Hare, the retiring Captain RNXS) and the new Captain RNXS (Captain John Griffiths) with rather a lot to do!

Moelfre

Moelfre, headquarters for RNXS exercises off Anglesey. Geminis are most useful when there are no piers or jetties available.

Facilities at Moelfre are sparse . . . . the local church hall doubles up as the Port Headquarters! Prayers and plans sometimes go hand-in-hand!

Chapter Seven

The Record of the Portsmouth and Medway Areas

XSV Loyal Factor*—briefly a Medway Command training craft—after less than a year she was re-deployed to Northern Ireland—and renamed HMS* Alert.

The decision to transform the RNMWS into the RNXS on the 5th November 1962 was to have major repercussions within the Portsmouth Command, which assumed responsibility for RNXS activities between Portland and Great Yarmouth, amalgamating the old "Portsmouth" and "Nore" Commands. The "New Portsmouth" Command was organised into six Port Areas—"EAST ANGLIA" (Gt Yarmouth, Lowestoft, Ipswich, Felixstowe and Clacton Units), "LONDON" (Maldon, Pitsea, Southend, Chelsea, Gateforth, Mortlake and Kingston Units), "MEDWAY" (Greenwich, Gravesend, Chatham and Sheerness Units), "DOVER" (Herne Bay, Thanet, Dover and Folkstone Units), "PORTSMOUTH" (Hastings, Eastbourne, Newhaven, Hove, West Sussex, Portsmouth, Isle of Wight, Southampton and New Forest Units) and "PORTLAND" (Bournemouth, Poole, Swanage and Portland)—a total of 33 Units with 1265 Volunteers. These were administered from the C in C Portsmouth's Offices (behind the Staff Officers Mess) by a staff consisting of the Command Naval Auxiliary Officer (CNXO)—Lieutenant Commander C L Sutton RN—assisted by three Administrative Officers, two at Portsmouth (Nick Carter and Bungy Williams) and one at Chatham (Jim Hotham), and by the RNXS Training Officer—(Lieutenant Commander John Ogle DSC RN and his Assistant, Chief Petty Officer Bailey) based at HMS Vernon. Each Port Area was supervised by a Port Naval Auxiliary Officer (PNXO) with either a Section Naval Auxiliary Officer (SNXO) or Chief Naval Auxiliaryman (CNX) in command of each RNXS Unit. Among the early PNXOs were PNXO E. Anglia (Ted Coult), PNXO Medway (Don Simmonds), PNXO Dover (Stan Willey) and PNXO Portland (John Feltham—best known as Admiral Mountbatten's Wartime Signal Bosun).

Early in 1963 the Southampton Unit moved into the old BOAC Aquila Air Base at 50 Berth in the docks where they remain to this day—despite a slow but continuing tilting of the building towards the water! The Dover Unit was moved into the Admiralty Pens, which had been the Naval Officer in Charge's Headquarters, with the MTB pens below. Keeping the accommodation dry and stopping leaks from the roof were to be persistent problems.

The Portsmouth Command RNXS was allocated three MFVs (1036, 1079 and 1257) for training, based respectively at Chatham, Portsmouth and Portland. In June 1963 at a PNXOs meeting at the Admiralty, the Vice Chief of Naval Staff (later as Admiral Sir John Frewen, to be C in C Naval Home Command) announced that the RNXS was to be equipped with 12 ex-Inshore Minesweepers over the ensuing 18 months, to replace the ageing MFVs. (It was at the same meeting that VCNS announced the impending appointment of the first Captain RNXS).

This was welcome news indeed, following the loss of MFV 1036 which had foundered on the rocks off the entrance to Dartmouth in thick fog at 2200 Thursday 25th April. A moving and detailed narrative by the late SNXO Bertie Mainprise, who was one of the crew, tells of the awful realisation that the ship was aground on rocks, position uncertain. The inability to pass distress signals over the TCS Radio. The scramble into lifebelts. The increasing list as she settled on the rocks at an angle of 40°. The silence, the darkness, the difficulty in moving about the ship. The cheer as Niton Radio eventually responded to the "Mayday" call. The relief when a voice hailing from out of the gloom, proved to be that of the Duty Lieutenant Commander from Britannia Naval College in a motor cutter. The appearance of two policemen in uniform on the rocks at Compass Point above the ship at daybreak, ready it seemed to hand out parking tickets! The hospitality of the College—the hot baths and the food! Finally, the editor's remarks over the ordeal of the Skipper—Mrs Jo Cook—at the hands of the Press—"she projected, both on TV and in the press a constant impression of dignity, loyalty and strength of purpose, and in doing so has created a most favourable impression both of herself and the Service to which she belongs".

In May 1964 four IMS entered service in the RNXS, including Thakeham (based at Chatham) and Woldingham (at Portsmouth). Later that year in Fareham creek the Woldingham was severely damaged, the engine and control rooms were flooded and the survey assed her as being beyond economical repair. She was replaced by the Portisham, which is still in service at Portsmouth, more than 30 years after she was launched at the Dorset Lake Shipyard at Poole. Autumn 1964 saw the first major NATO exercise in which the RNXS undertook their new role in headquarters ashore. Exercise TEAM WORK involved a convoy of eighteen merchantships. RNXS members manned the shore offices and 11 Boarding Vessels at Portland, Torbay and Mounts Bay.

1965 produced several changes in the Staff. The sudden death of Lt Cdr C L Sutton, on the eve of his retirement, saw the appointment of Lt Cdr John Ogle DSC as Command Naval Auxiliary Officer, Lt Cdr Eric Hill MVO to the newly titled post of "Naval Auxiliary Training Officer" as Head of the RNXS School in HMS Vernon and also the arrival of Lt Cdr Charles Beddoe as the Training Officer for the SE England Area covering some 20 RNXS Units, a post he held continuously until his retirement in December 1984.

Commander West—for many years in charge of the Littlehampton Unit.

In 1966, the Herne Bay Unit moved to Whitstable and was renamed Swale. Early in the year, a third XSV, the Shipham, joined the Thakeham and the Portisham in the Portsmouth Command and all the MFVs save one were phased out. The Portsmouth XSVs were soon spreading their wings with visits, to Amsterdam by XSV Thakeham from Medway, Duisburg (140 miles up the Rhine) and Honfleur by XSV Portisham from Portsmouth and to the Channel Islands and North Wales by XSV Shipham from Portland. During the visit to Honfleur, one of the Portisham's crew (Mr Trickett of Isle of Wight) rescued a Frenchman "who had imbibed too handsomely" and knocked himself out, falling off a quay into the harbour. The Skipper (Jo Cook again) reported that the first aid party applied artificial respiration to such good effect that when the ambulance arrived the rescued man was not only breathing but swearing too! The annual involvement began of the Medway XSV in the Mayor of Rochester's River Parade in which the Mayor exercised his titular role as Admiral of the River Medway. Lt Cdr Bill Harding was appointed as Assistant NXTO at the RNXS School and as Area Training Officer (South Coast). LNX Tom Stableford of the Sheerness Unit (later to be Head of Gravesend Unit and then PNXO Thames) qualified as a Skipper. George Daniel, who joined the RNMWS in 1957 at Newport South Wales, had transferred to Gt Yarmouth in 1959, where he became Head of Unit.

During the years 1963-69 Training Centres throughout the Command were experiencing problems. Some were in rented premises, a few in the homes of senior RNXS members, and others under sufferance in government-owned property. Training facilities were often inadequate, resulting in loss of members. Among the worst off were Pitsea, Southend, West Sussex (meeting in the basement of the local yacht club with a small shed in the car park to

RNXS Staff of the Medway Command—shortly before the Area was absorbed into the Portsmouth Command.

store training equipment!), New Forest, Bournemouth and Poole. However, 1966 saw some improvement with the moves of two London Units (Gateforth and Chelsea) into HMS Discovery with the Sea Scouts, followed early in 1967 by the Kingston and Mortlake Units. The Thanet Unit which trained at RAF Manston moved into their own quarters in Hawley Square, Margate, where they remain to this day as the Margate Unit. The Hove Unit moved, with the RNR Sussex Division, from the King Alfred Huts on the sea front at Brighton into a modern building at Maxwell's Wharf in the Basin at Shoreham.

In 1968 Captain A J Harvey MN was appointed PNXO Humber, a post he held until the 1986 staff restructure of the Humber Group when he retired after more than 22 years' service. In May, the Annual RNXS Church Service at All-Hallows-Next-the Tower (of London) was revived. It has become a source of spiritual guidance and reassurance for many members in these difficult times and is also a pleasant social occasion for both past and present members.

The first of the new class of Fleet Tenders were delivered in 1971. Their names were to be prefixed with the word "Loyal"—a happy reflection of the quality of the RNXS volunteers. A total of ten were planned as replacements for the eight IMS and delivery was to be phased over a number of years. 1972 saw the start of a major RNXS participation in Portsmouth and Chatham Navy Days. At Portsmouth over 3,000 passengers were taken for harbour trips, the money raised being donated to Naval charities. This has been a regular feature ever since and figures have escalated from 7,000 in 1975 to 19,000 in 1983.

The end of 1974 saw the appointment of Commander David Farquharson as ANXO and NXTO, while in 1975 Lt Cdr John Ogle RN retired as ANXO Medway and was relieved by Lt Cdr Duggie Baker. John Ogle was a serving Lieutenant Commander appointed to HMS Vernon in 1957 to be the RNMWS Training Officer. He joined the RNXS five years later and then became Command Naval Auxiliary Officer for the combined Portsmouth and Medway Commands in 1965 and Area Naval Auxiliary Officer to Flag Officer Medway in 1973, when the areas divided again. Another pioneer of the service also retired, PNXO Kent (Stan Willey), who had more than 22 years' service. In September the result of several years hard work came to fruition for PNXO Thames (Captain Tom Stableford) when the new training centre for the Gravesend Unit and Port Headquarters for NOIC London was opened by Flag Officer Medway (Rear Admiral Stephen Berthon) at 18 Building, Milton Barracks.

Tony Patten, Head of Hastings Unit, was appointed PNXO Dover in 1976 and was succeeded by his wife Judy whom he had first met at the Unit. The Pattens have become a by-word in the area, having given more than 40 years service between them. Also in 1976 Mr John Brodey,

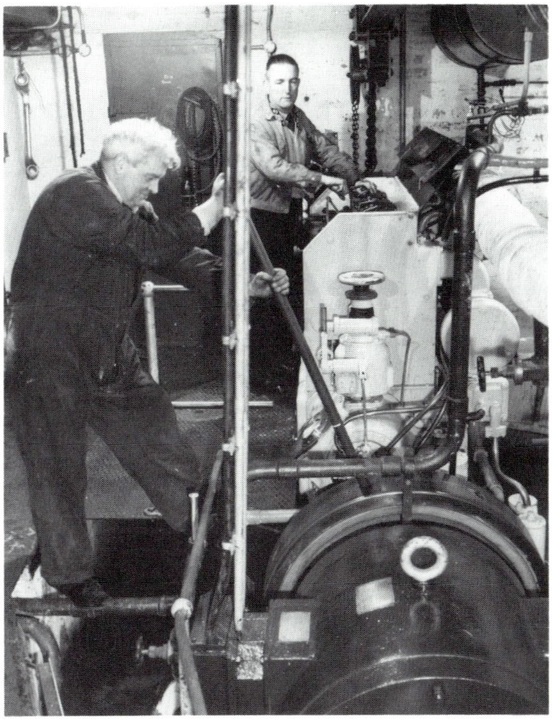

During the power strikes in the early 70s, the Dover Unit, lodged in the Admiralty Pens, successfully recommissioned the W.W.II generators to keep training going.

of the London Unit, was awarded the Queen's Commendation at a ceremony conducted by the Lord Lieutenant of Greater London, Lord Elworthy (Marshal of the RAF and formerly Chief of the Defence Staff) for saving a small child from drowning in the River Thames on the night of 26 August 1975 while serving as crew of MFV 1256. Mr Brodey jumped fully clothed into the river to rescue a small child who had fallen in and was being swept away on the strong ebb tide. Mr Brodey also received a Testimonial on Vellum from the Royal Humane Society. In a letter congratulating Mr Brodey the Commander in Chief, Naval Home Command (Admiral Lewin) described Mr Brodey's actions as bravery of the highest order. The child's mother remained quite unperturbed throughout an extremely tense twenty minutes, treating the matter as an everyday occurrence, as though her child might have fallen in the local duckpond. Ignorance is bliss indeed. In June the retirement of Mrs Jo Cook of the Portsmouth Unit saw the end of an era, for she had led the way in many fields—first woman skipper, first promoted to Officer's rank, first appointed Head of Unit. She joined the RNMWS in 1953, became Skipper in 1961 and was awarded the RNXS Medal in 1967 and the BEM in 1968.

1977 was a vintage year, including as it did the Silver Jubilee Fleet Review in which the RNXS played a major part. A flotilla of nine XSVs from all areas came together for seven days under the command of ATO Portsmouth (Lt Cdr Bill Harding) and carried out a variety of patrol and despatch tasks, culminating in the Review Ceremony by Her Majesty the Queen at which the RNXS carried out perimeter patrols, acting as barriers between the Royal Yacht and the armada of spectator craft. Those who were privileged to take part will never forget that day, especially Bill Harding on the eve of his retirement after 46 years in the RN and the RNXS.

Portsmouth Navy Days that year were immediately preceeded by the first Round the World Yacht Race starting from Portsmouth. For more than two weeks Vernon Creek was a haven for 15 yachts of all sizes preparing for this great adventure. The well known Robin Knox-Johnson, Eric Williams, Claire Francis and Eric Taberly were with us every day. On Saturday 27th August there was at Spithead one of the largest and perhaps the unruliest gathering of craft, large and small, sail and power, ever seen at the start of a Yacht Race. RNXS vessels tried valiantly to keep spectators clear of the contestants as they approached the starting line—but to no avail!

The retirement of PNXO East Anglia (Ted Coult) was marked by the award of Certificate of Meritorious Service to him in September by the Lord Lieutenant of Suffolk, the Earl of Stradbrooke. Ted was one of the very first RNMWS members and had played a major part in the early days. He was succeeded by George Daniel of the Gt Yarmouth Unit.

1978 began with the New Year Honours award of the MBE to the Head of Littlehampton (W. Sussex) Unit—SNXO Baron Nikolin. Both Area Administrative Officers, Jim Hotham from Chatham and Nick Carter from Portsmouth retired, having made major contributions to the RNMWS and the RNXS, each giving over 24 years service. Denise St Aubyn Hubbard became the second lady member to qualify as a Skipper in the RNXS. A Press and the TV day was arranged and it certainly sparked off major interest by the media. One devious journalist, seeking to provoke a display of chauvinism from a male member of her crew, was quietly put in his place by the reply "We judge our Skippers by whether we feel safe with them at sea—with Denise in charge, we sleep well in our bunks at night"!

In September the three Units of the Portland Group (Poole, Portland and Bournemouth) transferred from FO Plymouth to FO Portsmouth—thereby reversing a previous move in January 1975—originally made necessary by changes in regional boundaries. This had in turn reversed an earlier move from Portsmouth to Plymouth in 1969 when the Naval Home Commands were re-organised! On the night of 12 December a terrific storm completely cut off Portland from the mainland, with winds up to hurricane force, judged by many to have been the worst in living memory and comparable with the legendary storm of 1824 when more than 20 people were

The Jubilee Fleet Review 1977

All the RNXS vessels took part in the Review.

Keeping the large number of spectator craft clear of the channels was a major task of the RNXS on the day of the Review.

The craft looked their best

and did a hundred and one other tasks throughout the week.

A Policeman was embarked in each craft on patrol duties. Lacking "powers of arrest", the Service might have needed this authority to enforce law and order in the waters of the Solent.

drowned. Fortunately there was no loss of life this time but extensive damage was caused along the whole length of Chesil Beach. Perhaps such events help to explain the independence and resilience displayed by Portland members over the years.

In January 1979 XSV *Loyal Mediator*, the last of 5 RNXS fleet tenders built by Richard Dunstan Ltd of Thorne on the Humber, arrived at Portsmouth. This coincided with a move of the Portsmouth Unit within HMS *Vernon*, with close on 100 members, from 7th Floor Creasy Centre to spacious quarters in 17A Building, providing ample training spaces for all four branches.

The funeral of PNXO J Feltham MBE, one of the early pioneers of the RNXS at Portland, took place in Weymouth in July 1980 and his ashes were committed on the waters in the Bay. Soon after, we lost a much liked versatile and respected character of the Dover Unit, Mr Bertie Mainprise—Skipper, Charge Engineer, Communicator and finally, Head of the Unit.

Chief Naval Auxiliaryman Dickie Dawes of the Southampton Unit was presented with the British Empire Medal by the Lord Lieutenant of Hampshire, the Earl of Malmesbury, at a ceremony in the Guildhall at Winchester on 26 June 1981. Dickie's award was as a result of his excellent work and training on communications. The news that the Chatham Naval Base would be closed and that FO Medway would disappear in 1983 was a great shock to everyone that summer. For administrative and operational reasons, it was arranged that the 16 Units from Shoreham to Great Yarmouth would join the Portsmouth Area and the 3 Humber Units transfer to the Scottish Area, with XSV *Shipham* deployed to Portland and *Loyal Helper* to Rosyth.

1982 is remembered for the Falkland Islands War and the never to be forgotten departure of the Task Force on that Monday in early April from Portsmouth Harbour. Although offers of support were made by the RNXS during this tense period there was little that the Service could do. However the *Portisham* escorted and welcomed the ships on return from their successful mission. Here is a report written at the time:-

> "It was our privilege to be there when the heroes returned to a welcome which matched the scenes at the end of the Second World War. The *Portisham* provided a marvellous platform from which to join the armada of small vessels when the *Queen Elizabeth II* returned with survivors from HM Ships *Sheffield, Coventry, Ardent* and *Antelope*. The Royal Yacht with the Queen Mother on board was anchored in the West Solent as *QE2* with the 1000 survivors on board steamed past in salute and then entered Southampton Water to berth in the Docks. The scene was repeated when the *Canberra* with 3000 Royal Marines and soldiers arrived to another tumultuous welcome. *Portisham* was right there escorting both ships all the way from Cowes to their berths alongside. Piloting an IMS surrounded by literally hundreds of wildly enthusiastic yachts, speed boats, dinghies and many other assorted craft was hair raising and exciting, but the experienced skipper—CNX Sam Dobell—proved a match for them all and we returned safely after two truly exhilarating experiences."

Navy Days at Portsmouth that summer were renamed Naval Base Open Days, because it was not possible to open HM Ships to the public so soon after their return from the South Atlantic. Although naturally the attendance figures were down (63,000 compared to 90,000 the previous year) the crowds were as enthusiastic as ever and the decision not to cancel the event altogether was fully justified. Sensing that RNXS harbour trips might be more popular than usual, five XSVs were assembled instead of the four of previous years. The *Shipham* and *Loyal Helper* came from Medway, *Loyal Mediator* from Portland, *Loyal Chancellor* from Plymouth and the *Portisham* from Portsmouth. All five vessels were kept going flat out throughout the three days. Fortunately the weather was kind and all our records were broken. Nearly 19,000 passengers were taken for a full half hour tour of the ships berthed on the tidal walls of the Base and nearly £8,000 was raised for Naval Charities. The splendid performance of the crews and the untiring and efficient support of the ticket sellers, marshals, berthing parties and communicators were responsible for this success, making a major contribution to the Open Days and providing support for the Navy when needed.

Later in the year the "Tall Ships Parade of Sail" assembled at the Ocean Terminal, Southampton. Their procession down Southampton water, escorted by the *Portisham*, was a most colourful event.

With the closure of the Medway Area in 1983, a reorganisation of the Portsmouth staff began with the appointment of Commander John Leathes as Area Training Officer. Having been the first Commander of the Maritime Trade Faculty, his experience was to be of great importance particularly when, in 1986, he became Naval Auxiliary Training Officer.

In October, the IOW Unit was chosen by Admiral Sir Desmond Cassidi for the first routine visit to an RNXS Unit by the Commander-in-Chief. He took passage to Cowes in the *Portisham* and spent some two hours inspecting members under training, afterwards addressing the Unit. This set the pattern for subsequent visits by CINCNAVHOME to Units throughout the United Kingdom. In the same month, the merger of the Bournemouth Unit with the Poole Unit at the RM Camp at Hamworthy also involved the retirement of two long serving Heads of Units, Marsie Taylor, BEM, of Poole, and Roy Hill of Bournemouth, with 20 and 23 years service respectively. Roy and his wife (both in their early 60s) emigrated to the United States of America where they now have a thriving restaurant business. The closure of the Bournemouth Unit after 30 years was a sad occasion. A good and proud Unit fell through changed requirements. At the farewell Reception after his valedictory speech, the ANXO was presented with a commemorative gift, a small model axe embedded in a stump inscribed "Woodman Spare That Tree"!

The Maldon Unit, which had operated under trying conditions in several temporary premises finally found a warm and spacious home in a senior army officer's married quarter on the outskirts of Chelmsford, whence it took its new name. The Chatham Unit, which had moved 7 years earlier from some well-loved huts in HMS Pembroke to a splendid training centre in Collingwood Block in Khyber Road nearby, suddenly found their privacy invaded by the RNR, RMR and Cadets following the closure of Pembroke. From being the sole occupants in the West Wing, they had become one of five lodger Units in the new HMS Wildfire.

By 1984 the assimilation of 16 RNXS Units (ex Medway) had increased the workload and stretched even the re-organised and augmented staff at Portsmouth. With only 3 Training Vessels (*Portisham* and *Loyal Mediator* at Portsmouth and *Shipham* at Portland), arranging adequate afloat training became very difficult. The news that four new P20 Coastal Training Craft, capable of 24 knots, were to be provided for the RNXS was therefore welcome, particularly as the first two, *Example* and *Explorer*, were to be allocated to Portsmouth.

The RNXS Staff re-organisation began to affect the Portsmouth Area in 1985 with the appointment of Group Naval Auxiliary Officer East Anglia (Lt Commander Peter Brooke) followed by GNXO Thames (Lt Commander Duggie Baker) and (in 1986) GNXO Dover (Lt Commander George Claydon) as the old Group Training Officers were phased out. The Lowestoft Unit, which for many years

The RNXS is proud of its lady skippers. Denise Hubbard is a yacht delivery skipper and sailing instructor in civil life—and frequently commands RNXS craft.

had struggled along uncomplaining on the 4th floor of an old building under cramped conditions, were rewarded for their patience. A former coastguard officer's house was extended and converted into a spacious, well-equipped training centre which was officially opened by Captain RNXS (Captain J R Griffiths, RN).

The long delayed move of the Area Staff from HMS Vernon into Semaphore Tower in the Naval Base took place in 1986, not without the usual dramas such as having no furniture and no heating. The move, combined with staff changes, the arrival of the new Training Craft, and planning for our Silver Jubilee Year generated much paperwork and provided a challenging prospect, now largely overcome.

Over the 25 years since the RNXS was founded, the majority of the original Units within the Portsmouth Command have been retained, some merged with other Units, and most have moved into more spacious quarters with better training facilities. New units have been opened and inevitably several were closed. In 1962 we started with 33 Units and 1,265 ex-Minewatchers—today we have 23 Units with around 1,100 Naval Auxiliarymen. Faces and names may have changed but the devotion, the loyalty and, above all, the sense of humour continue to shine like a beacon of hope.

Awards

Prominent Portsmouth Area RNXS personnel are presented with their RNXS medals—after achieving 12 years very good service (or undetected crime!).

Chapter Eight

The Record of the Plymouth Area

1963—an inter-command minewatching competition at Dartmouth.

There were 32 Units with a total strength of 1,342 members in 1962, Commander Richard Gabbett-Mulhallen had just taken charge of the area and with the chief concentrations of merchant naval activity on west coasts at that time, Units were not slow in adapting to their new role. Not that Plymouth area had been backward in minewatching either: the inter-command competition had been won twice despite the ''unfair'' emphasis of minewatching exercises organised for the Nore and Portsmouth commands where, it has to be admitted, some of the best mining areas lie. During the 1960s the re-activation of Naval Control of Shipping as an ongoing requirement was led by the Naval Staff at Plymouth, all the major exercises at first being scheduled in the Plymouth area. To get NCS training into ports, the Plymouth Command began the Devon Coast series of local exercises in which RNXS members played a large part and these later spread to other Commands. The first live NATO exercise in which the RNXS were involved, codenamed ''Skuber-Mor'', took place in Mounts Bay in May 1963. From a temporary office set up in the Ship Institute in Newlyn, a convoy of 40 ships was received and sailed with each being given the necessary sailing orders. Members manned the office and two RNXS MFVs were used for boarding and berthing. It was very tiring work and members had little sleep during the four days of the exercise. It was a trial run in many ways, to train personnel in NCS procedures which were found to be cumbersome and time consuming. Everyone regardless of rank was pressed into writing up sailing orders and the NCS office ran out of stationery, so the Administrative Officer, Lieutenant C O Sadler and the Head of the Unit, Maurice Hogg, were despatched to the local shops to buy more! The MFVs performed valiant service but were slow and could not get round all the 40 ships of the convoy. These were in fact not merchant ships but NATO minesweepers so they were able to provide the necessary degree of self help.

During the 1960s, Units, their strengths and their locations, were adjusting to the new role; there were a large number of busy west coast ports and much recruitment of members was necessary. By 1966 the area strength had risen to 1,663 members. In the Bristol

A Large Area

From the North . . . . an early (1959) group photograph of the Barrow-in-Furness Unit

. . . . to the deep south. Jersey Unit pose for the camera during the visit of the Lieutenant Governor of Jersey (Admiral Sir William Pillar, KCB) to declare their training centre open.

The Guernsey Unit pause for a breather, and photograph, during a Channel Island exercise.

Channel area, the Bridgwater Unit had closed and was re-formed at Burnham-on-Sea; the Clevedon Unit was going strongly in support of Bristol. The Sharpness Unit was fading and was soon to be re-established at Gloucester. In West Wales the Fishguard Unit joined with the Milford Haven Unit in the old Flag Officer's headquarters. Later, when the Neyland Bridge was eventually completed it seemed sensible to concentrate the Milford Haven and Pembroke Dock Units with the Resident Naval Officer in the naval yard. Further along the south coast of Wales, there was little change in the Units until the seventies when numbers were allowed to run down to match the reduced loading at the ports. Swansea, always a strong Unit, was often used in support of Milford Haven and Barry, both of which had major co-ordinating roles for Bristol Channel area. Mr Ivor Swarts was made the Head of Barry Unit in 1962, the youngest by far, and he remained in charge at Barry until 1985, a splendid and unbeaten record of service. The Cardiff and Newport Units maintained a steady strength, participating in numerous exercises until the decline of the port of Cardiff and the continuing importance of Barry forced a merger of the two latter units at Sully, hard by Barry. Mr Edgar Coldwell was appointed head of the Newport Unit in 1964 and still serves today—another remarkable record—while Mr John Dukes became head of Cardiff in 1971, was appointed Port Naval Auxiliary Officer for South Wales in 1978 and remained until 1987 when he became Staff PNXO at Milford Haven. There must be something about the South Wales coast that produces such longevity and service!

To the north in the Irish Sea area, the Anglesey Units of Holyhead, now under Mr John Cave and Menai Bridge, under Mr D East, survived accommodation problems with remarkable aplomb and efficiency and provided a scratch Port Headquarters in the Church Hall at Moelfre which remains a tribute to the British genius for improvisation. Perhaps, again, continuity is partly the reason; for Captain Lawrence Williams, Royal Marines, served as the Port Naval Auxiliary Officer for the area from 1964 to1987. A Unit was formed in Llandudno in 1974 to give support to Moelfre, which is now situated in nearby Conwy in better accommodation. The Birkenhead Unit at Morpeth Dock was and is the largest in the RNXS; it is also the best equipped and probably the most social Unit of all. From discos and dances to dinners, their social calendar always seems full but they have worked hard too, supporting innumerable exercises and afloat ventures. Maybe big is beautiful; anyway, coupled with the Mersey sound it has advantages! A little later than some, Preston Unit was formed in 1955 in response to the requirements there and at Heysham; presided over by the Wilson family, it has been a most effective support Unit. The Unit at Fleetwood, although small, has been rock steady in an area of much change and is now on the up and up.

Mr Douglas Bulloch has been an important leader for the north-western area. Joining the Service as an Admiralty Ferry Crew Skipper in 1963, his expertise gained him rapid promotion to Head of Unit and in 1978 he was appointed Port Naval Auxiliary Officer for the Mersey Area. In 1986 having retired as a schoolmaster, his 'proper job', he was appointed temporarily to the permanent staff as the

Torrey Canyon

The Torrey Canyon *disaster. Many long hours were spent spraying the oil from makeshift booms rigged in RNXS inshore minesweepers.*

The "Command Team"—Cdr Task Group 312.3 (second from right) and his staff.

Group Naval Auxiliary Officer on the sudden and tragic death of Lieutenant Commander John Hodgson. The RNXS is fortunate indeed to have the benefit of such continuity and expertise. The Unit at Barrow-in-Furness closed in 1979. It had given good professional service over the preceding years and enjoyed the support of Vickers Ltd who provided the majority of members. However, the requirement for a Unit at Barrow had declined and it was inappropriate to make it a Support Unit. Mr L E Dribbell had been the much respected Head of Unit. The small Whitehaven Unit, despite accommodation problems, kept going and through the efforts of Mr Frank Pearce, another long serving Head of Unit, became the most qualified per capita in the Plymouth area. The Unit moved to Workington, the main port and finally to the NATO Armament Depot at Broughton Moor nearby, where Mr D Futers is now in charge.

Having reviewed the progress of some of the Units over the years, we return to the Summer of 1965, when the west country Units began supporting the Naval Air Command Sub-Aqua Club (NACSAC). Members of this club had visited the Scilly Isles the previous year looking for the wreck of HMS *Association*, the flagship of Sir Cloudesley Shovel. The *Puttenham*, with an RNXS crew skippered by Lieutenant Commander Ted Barter, spent some 6 weeks supporting 16 divers searching off the Western Rocks: although some material was recovered they were not successful, bad weather aggravated by the treacherous tidal conditions having greatly reduced the diving time. In the Summer of 1966, the *Puttenham* with 24 divers returned but they were unable to develop their full potential due to bad weather again, however they were able to conduct an alternative survey of crawfish in more sheltered waters.

Following the discovery of a 1765 chart and a resulting reassessment, the Summer 1967 expedition decided to search in a different area. On their third day after arrival, twenty cannons were located and a gold coin recovered. Early attempts to recover a cannon with makeshift equipment failed, then bad weather once more intervened. A few days later with the assistance of a press boat and the use of aircraft recovery air bags, the *Puttenham* recovered a large bronze cannon, well preserved and with Duc de Beaufort's coat of arms, later surrendered to the Receiver of Wrecks. This cannon may now be seen in the Tower of London. Media interest was intense and although MOD granted an extension of time for the expedition, diving became more dangerous. Commercial interests were granted rights to search the area and regrettably, in the frantic search for the "great treasure", explosives were used. However, the NACSAC expedition did discover several more cannon and a huge bower anchor. In an effort to identify the wreck one more expedition by NACSAC was mounted a month later in September with XSV *Odiham*. No spectacular finds were made but a host of coin and small domestic items were recovered, one being a gold wedding ring engraved "God above increase our love". Although the great treasure was found later by commercial interests and the wreck which the NACSAC expeditions had located was not identified as HMS *Association*, nevertheless they had positively located the wrecked fleet and provided evidence to which the RNXS vessels had made a useful and not unimportant contribution.

The following year, 1968, the *Puttenham* again worked with NACSAC off the Scilly Isles, but diving on HMS *Association* was only part of the programme.

The area's effort was much in demand in the late 1960s for there was at least one major and one minor exercise a year to contend with and also a large basic training backlog to overcome. While the summer NACSAC expeditions were continuing, there was a further call on the RNXS when an emergency arose on 18 March 1967 with the grounding of the tanker *Torrey Canyon*, fully laden with 110,000 tons of crude oil for Milford Haven, on the Seven Stones reef. Refloating was impracticable and despite attempts to set fire to the oil and to contain it, huge oil slicks menaced the Cornish beaches and wildlife. Every available naval craft capable of spraying was pressed into service, organised into two Task Groups based on Falmouth and Newlyn. Commander Gabbett-Mulhallen was sent to Falmouth in charge where he set up a headquarters in the dockyard for the group of nine vessels operating from there. These included four RNXS: the *Puttenham, Shipham, Odiham* and *MFV93*, equipped for spraying with booms, pipes, pumps and drums of detergent. The crews had interesting and exhausting experiences, working from dawn to dusk in all weathers, at times overwhelmed by the slicks, suffering at first from shortage of detergent, then cursing the detergent for making their decks so slippery and dangerous! At the end of the first week there were crew shortages in the RNXS vessels, particularly communicators, and volunteers were sought from Wales and the North. Meanwhile RN personnel were drafted in. By the 1st April, air surveillance showed a much improved situation and the operation began to wind down. Throughout, the two Port Headquarters were given valuable support by the Falmouth and Penzance Units, who kept the plots and records and assisted with communications. As ships and aircraft of the fleet were involved in the operation, RN personnel and equipment were used for communications. The four RNXS vessels were released on 5th April to return to routine training duties.

During the early 1970s, the accent was on basic training in the Units and on trying to improve accommodation, although the 'across the board' cuts in expenditure made the latter a slow process. While the committee reviewing the RNXS in 1974 gave the Plymouth area a clean bill of health, reductions through 'wastage' continued. Then in 1976 the Hicks-Beach report reduced the numbers and size of Port Headquarters, particularly on the western coasts, so that by 1979 the RNXS area strength was down to 850 members. Even in the south there was contraction; the

An early West Country exercise. C-in-C Plymouth (Admiral Sir Charles Mills) inspects an MFV's crew.

Still smiling, despite their cramped training centre, the local Unit at Menai Bridge.

The fellows from Fleetwood—provide crews for their training vessel based on the Mersey.

Cdr Tom Trick (ANXO Plymouth) inspects the piping party.

Sir Harold Wilson (and Paddy) pay an informal visit to XSV Loyal Moderator *at St Mary's, Isles of Scilly.*

Salcombe, Kingsbridge and Dartmouth Units all closed down during the seventies. Torquay and Paignton Units reformed at Brixham in the Customs House; Teignmouth Unit withdrew into Exeter and the Exmouth Unit after many years at the Yacht Club, sought permanent accommodation with the Royal Marines at Lympstone. On the North Devon coast, the Ilfracombe Unit had closed. The Barnstaple Unit moved into Fremington Camp and gained much from the local support. Burnham on Sea joined with the Clevedon Unit and in turn, Clevedon joined with the large Bristol Unit. The Portland, Poole and Bournemouth Units were transferred to the Plymouth Area on 1st January 1975 as a result of an uncomfortable boundary adjustment. Always independent, the hearts of these splendid Units somehow remained with Portsmouth and so they returned four years later despite the boundary. Perhaps it was just a question of so near and so far. After a long gestation period, the Guernsey Unit was formed in 1974 and quickly gained strength and expertise under the leadership of Lieutenant Commander Charles Poynder. While there were bright spots, it was a matter of much sadness that so many good Units had to amalgamate or close in the 1970s. However, the remaining Units became steadily more proficient as training improved and as the yearly round of NATO and national exercises increased in complexity and demanded regular support. Commander Richard Gabbett-Mulhallen, the Area Naval Auxiliary Officer, had to retire due to ill health in 1975 and was relieved by Commander Tom Trick, the Training Officer in South Wales, who in turn was replaced by Lieutenant Commander John Styles.

The RNXS became better known through publicity and visits. Amongst the highlights were the visit of Prince Charles to the Barry Unit in May 1975 while his ship HMS *Bronington* was visiting the port and an informal visit by the then Prime Minister, Sir Harold Wilson, with his dog Paddie to XSV *Loyal Moderator* while they were on holiday at St Mary's in the Isles of Scilly. For the Silver Jubilee Review at Spithead in 1977 all three craft—*Loyal Moderator, Loyal Chancellor* and the *Puttenham*—attended that great event and afterwards paid visits to Channel ports before returning to their bases. In 1978 came another tanker emergency. The *Amoco Cadiz* suffered a steering breakdown and was driven ashore in heavy weather near Ushant. Tons of oil streamed eastward along the French coast towards the Channel Isles. The Guernsey Unit was ordered to activate the Port Headquarters: it was operating in just four hours. Three RNXS craft, *Loyal Moderator, Loyal Chancellor* and *Loyal Mediator* from Portsmouth were manned and put at short notice to assist with oil dispersal in the Channel Isles, while other specialist tugs and auxiliaries were sailed to Guernsey. The North Westerly gales kept the oil away from the Channel Isles but not the French beaches. The danger passed for the Channel Isles and after five days the Port Headquarters, which had proved a valuable communication link, was stood down and the craft were released.

The Guernsey Unit was again in the news in June 1978

The Commanding Officer of HMS Bronington *visited the Barry Unit in May 1975.*

The Head of the Unit—Ivor Swartz (one of the longest serving Heads of Unit) explains an ''historical'' minewatching sight to His Royal Highness.

when they were inspected by Her Majesty the Queen during her Jubilee visit to the Island. The Unit was the only auxiliary or reserve of the Armed Forces in being in the Channel Islands.

As we entered the 1980s there were a number of staff and Unit changes. In 1980 the Area Naval Auxiliary Officer, Commander Tom Trick, retired and was relieved by Commander Brian Prideaux from the staff of Flag Officer, Plymouth. Lieutenant Commander Bill Bethell retired after long and loyal service in the Mersey Area. Starting as the Administrative Officer for the Northern Area, he took on the onerous task of Group Training Officer in 1973 and even now, having retired, still lends a hand when things are busy. In 1981 Lieutenant Commander R J Griffin (Griff), the tireless and dedicated Admin. Officer for the Plymouth Area, retired after some fifteen years' service. He was presented with a silver tankard on leaving which had a very suitable device on the lid, a Griffin—a ''fabulous creature''! Lieutenant Commander Sandy Powell retired as PNXO Devon after eighteen valuable years. His great contribution was to afloat training which he organised for his area, setting and maintaining high standards during the early RNXS days.

Recent Unit changes have been primarily to improve accommodation and facilities or to build up Unit strengths. Cardiff and Barry Units joined together at HMS Cambria, the Port Headquarters and RNR training centre at Sully. Exmouth Unit from Lympstone joined Exeter and the combined Unit has now moved to new accommodation in Wyvern Barracks, Exeter, while in North Wales, the Llandudno Unit became Conwy Unit in 'a cold store' and although under notice to quit, remains active—

The old training centre at Birkenhead with XSV Puttenham *alongside in Morpeth dock.*

If you've got a complaint—tell the Admiral . . . . Scrubbing these decks ruins your hands!

Life's tough at the top!—Lt Cdr Sandy Powell with some of his crew.

members' circulation must be good. The Anglesey Units, which had worked as one for many years, separated into the Holyhead and Menai Bridge Units, Holyhead recently moving into good premises after suffering for so long with a leaky roof; Menai Bridge's accommodation, having held the distinction of being the smallest training centre in the Service, was at last extended. Liverpool moved from their large and loveable old go-down shared with the Royal Marine Reserve at Morpeth Dock into the fine South Atlantic Building, with modern and improved facilities, at Vittoria Dock.

The Swansea Unit moved into spacious accommodation at the Old Custom House and were joined by the RNR CTC under the Navy House concept to become HMS Dragon. The Falmouth Unit moved from their wartime huts at the Hornwork back into town in more commodious and convenient huts at Bar Road. Finally the Plymouth Unit, after 21 years in their centre at the old gunnery school down by the jetty in HMS Drake, moved to a new block in Morice Yard in the summer of 1987.

So after a lengthy period of lagging behind the appropriate standards, the accommodation of Units in the Plymouth Area has taken a big step forward. Our strength has increased to 980 members as the build-up proceeds to meet our additional responsibilities in defence of ports and anchorages. We are now in the middle of the re-organisation of the permanent staff, which introduced the Group Naval Auxiliary Officers, to provide a higher degree of service and training for members and new fast craft are expected shortly. So there is plenty of interest and much to do. As the Flag Officer, Plymouth, said recently to the Holyhead Unit when opening their new training centre "The cuts in the Royal Navy over the next few years will mean we will rely even more on our Reserves and Auxiliaries for support".

Chapter Nine

The Record of the Scotland and Northern Ireland Areas

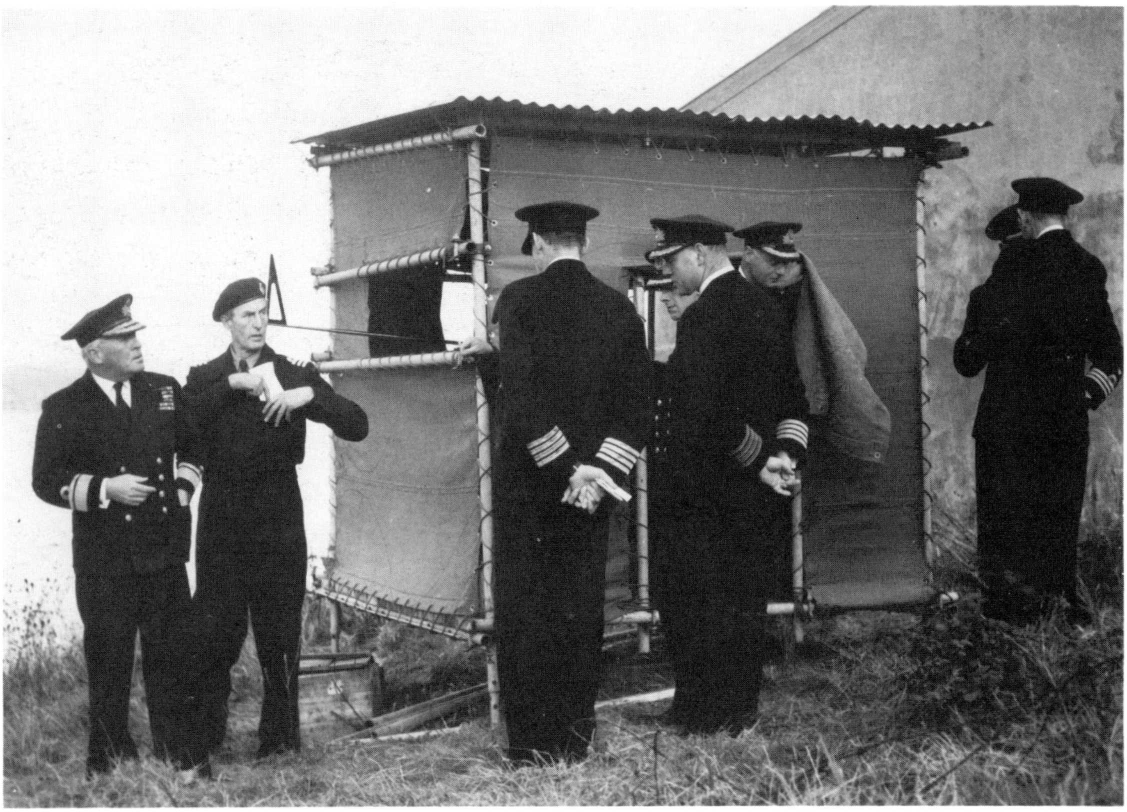

Who's next? Not what you might think, but Navy Works' answer to the request for weather protection for the minewatcher and his sight. This new construction clearly needed Senior Officer's approval (October 1953).

North of Edinburgh and spanning the Firth of Forth are two of the World's Engineering triumphs:- the Forth Rail and Road Bridges. Close to the northern end of the bridges and in the Kingdom of Fife lies the Maritime Headquarters (MHQ) of the Flag Officer Scotland and Northern Ireland (FOSNI). Here at Pitreavie, maritime affairs in the Northern Region are controlled and the Area staff of the Royal Naval Auxiliary Service is located. In 1949 the Admiralty had issued the directive to the Naval Commands to study the "Correlation of Minewatching, Plotting, Location and Destruction of Mines dropped from aircraft". Then a notice appeared in the press in February 1952: "It is intended to establish Minewatching Posts on the Clyde, at Aberdeen, Montrose, and in the Firth of Forth. The Minewatching Service is open to men over 45 and women over 26 who live in coastal areas. It is fitting that we should salute this new service as her Majesty the Queen has recently approved a specially designed uniform with appropriate signia". This was the beginning of the Scottish area involvement, the Service was established, Lieutenant Commander G B Hedley RNR was appointed as the Command Minewatching Service Officer (CMWSO) and his deputy was Commander B Dean DSO, RN Rtd (DCMWSO).

The Inverness Unit in 1959—operated from a garden shed near the PNXO's house.

The following reports give the flavour of "live" minewatching training:

"(1) Members from the Gareloch and Greenock Units took part in an exercise at Rosneath on Sunday, in which three Minewatching Posts, Logging Centre and Plot were manned. Mine splashes were simulated by red Verey's lights fired from an MFV which made four runs. Some difficulty was experienced at the Plot through small errors in the alignment of one or two instruments but when corrections for these had been applied, good fixes were obtained".

"(2) The helicopter, piloted by Lieutenant Lines RN, landed on the temporary heliport at North Kessock where a party of the Inverness Sea Cadet Corps loaded 6 'mines (40 lbs of shingle tied loosely in red net carrot bags) and the aircraft took off for its first run. A rapid lay of 6 mines was made from 1,000 feet from east to west in Kessock roads. The "mines" were most successful, the bags burst with a loud bang on hitting the water and raised splashes some 10 feet high."

By February 1961 Minewatching Units had been established at Inverness, Aberdeen, and on the Clyde at Ardrossan, Dunoon, Glasgow, Greenock and Cambeltown. In the Forth Area were Units at Edinburgh and Rosyth, and on the Humber at Hull, Grimsby and Goole. The Tees area comprised the Hartlepool, Middlesborough, Darlington and Saltburn Units and on the Tyne, the Newcastle and Tynemouth Units. Those founder members did a superb job in setting up what was to be the core of the Royal Naval Auxiliary Service. Early in 1962 Lieutenant Commander Hedley RNR handed over to Commander C E Emerson, OBE, DSC, RN who was to spend the next 15 years shaping the newly forming RNXS in Scotland and Northern Ireland. The formation was announced in the Scottish national and local press:- "The protection and control of our shipping in wartime is a vital task in which you can help. Men and women over 21 years of age may join the Royal Naval Auxiliary Service. Uniform is provided and expenses paid". The RNXS thus assumed a different role from its predecessor. The volunteers, however, never changed and the members accepted the new challenge with enthusiasm.

Early in 1962 the Invergordon Unit was launched and the first efforts were made to recruit in the Orkney and Shetland Islands. Mr B M Watt was appointed PNXO for North Scotland. Brian had built up the Inverness Unit from scratch in 1956 and the aura of those early days continues even today at his home at North Kessock. A Naval boundary change transferred the Northern Ireland area from C-in-C Plymouth to Flag Office Scotland and Mr Alec Paterson was appointed PNXO for Northern Ireland.

In 1963, MFV 175 was allocated by Captain-in-Charge Clyde and organised afloat training in the Clyde area began. A St Andrews Day parade was held in Glasgow where, for the first time, an RNXS contingent was mustered and inspected by Vice Admiral A R Hezlet KBE, CB, DSO, DSC (FOSNI). In the Forth Area the Rosyth Unit was installed in its own splendid accommodation at HMS Safeguard, the Naval Boom Defence establishment, and is still there today.

1964 saw an upsurge in the Afloat Branch. Early in the year, MFV 110 had been allocated to the Northern Ireland Area and was joined by the newly acquired Inshore Minesweepers *Birdham* and *Tongham*. These three Auxilary Service Vessels (XSVs) served the Clyde, Northern Ireland and North West Scottish areas for many years. XSV *PAGHAM* joined in 1965 to fulfil the training requirements on the East Coast from the Forth to the Humber with MFV 65 and the following year, MFV 105 replaced MFV 65 which redeployed to the Clyde. The priority task for the afloat branch at this time was to train crews to meet the war role at Assembly Anchorages but other tasks were undertaken and XSVs were kept busy throughout the summer. For example, in 1965 the *Birdham* acted as guard ship for Clyde Week Regatta at Largs in early July and then continued on a summer cruise to the Western Isles with a Stranraer crew. The *Tongham* spent a month in Northern Ireland from mid-June on local training and then moved to the Clyde Area in time to act as guard ship at Dunstaffnage Bay for the Sea Scouts Rally organised by Captain G Pound, CB, DCS, now Sea Scout Commissioner for Scotland. Returning to the Clyde, both craft were prepared for the Royal Review. The Clyde Port Auxiliary Service made a most efficient and speedy job of repairing the *Birdham*, which had sustained propeller damage due to not looking where she was going and the Clyde Units worked like beavers on both craft with paint and polish, bringing them up to the necessary standard of smartness for the first RNXS participation in a Royal Occasion, Her Majesty's Review of the Home Fleet at the Tail of the Bank.

On the East Coast in August the *Pagham* completed her adaptation, which included the removal of her minesweeping generators to make more training accommodation available, and sailed for Hartlepool with a Tyne/Tees crew. A busy fortnight was spent in Tyne/Tees area but unfortunately, when she was due to leave for Grimsby in September, easterly gales kept her weather bound in Hartlepool and a long planned programme to Grimsby had to be cancelled. This was not the first nor the last time that the long suffering Humber afloat sections were to be disappointed. The following week the *Pagham* made a safe passage to Rosyth and thence proceeded to the Tay to continue training.

With this degree of afloat activity, the shore elements may have seemed quiet, but a fair number of exercises were conducted. ''Dogwatches'' at Rosyth attracted good turn-outs and the Clyde Units were kept busy with their weekend exercises. In addition they contributed to a ''Meet the Navy'' exhibition in Glasgow and to a ''Royal Marine'' display at Greenock.

1966 saw the opening of the new Training Centre for Aberdeen in the old lifeboat house. Lieutenant Dennis Hedgecock was appointed as PNXO for the Tyne Area, relieving Lieutenant Commander Eccleston. Mr Tom Cramond became the Head of the new Unit in Stranraer which had moved into its tiny Training Centre in Stranraer Castle the year before.

Grandma's afloat . . . . members of the Rosyth Unit remember with affection Rosa Barker—stalwart of the Unit, here serving in the Loyal Volunteer.

1967 saw the appointments of Lieut Fred Hull as the Administrative Officer vice John Davidson BEM and Lieutenant Commander Erlend Clouston, DSC, RN, as the Area Training Officer for the Clyde and NW Area, replacing Captain D Adey Merchant Navy. Advance level training gathered momentum, the first Chief Naval Auxiliaryman (CNX) course at Rosyth being conducted. In August Chief ERA Arthur Morrell became the Area Engineer Officer (AEO). He had been drafted on loan from the Royal Navy soon after the arrival of the Inshore Minesweepers and now transferred to the RNXS permanent staff.

In July XSV *Birdham* suffered a fire at sea, always a dangerous matter, particularly in a wooden vessel. She was returning from a successful training cruise to Irish Ports when a serious electrical fire occurred off Loch Ranza. Fortunately no one was injured; SNXO Crosbie was the Skipper and his fire-fighting crew did some fine work under difficult conditions in getting the fire under

Thurso Unit are presented with a prize in the efficiency shield competition.

The new RNXS Training Centre at Ayr is opened by Mr George Younger MP (later to become Secretary of State for Defence).

The Inverness Unit muster onboard HMS Albion *at Invergordon for the presentation of the RNXS medal to their SNXO by Captain Henry Leach (later First Sea Lord).*

XSV Loyal Proctor *represents a damaged tanker during "Exercise Opener" off Sunderland. One of a series of "disaster at sea" exercises.*

control but not before considerable damage had been done to the engine room. Investigation showed that the fire was caused by a tight cable clip chafing through the insulation of one of the main power cables in the engine room, where they were subjected to considerable vibration. The Board of Enquiry found that although the *Birdham* was severely damaged, the cause of the fire was in no way attributable to the crew and the conduct of the Skipper and his crew in getting the fire under control, disembarking surplus personnel to safety and getting the craft back to Greenock, was a thoroughly professional performance in keeping with the highest traditions of the Royal Navy. Mr Crosbie was subsequently awarded the MBE.

In 1968 a minor staff re-organisation took place. Erlend Clouston moved to Inverness as the Area Training Officer covering North Scotland and Northern Islands and Lieutenant Commander Harry Fowler became ATO South Scotland. A new Unit at Stornoway was established under Lieutenant 'Ernie' Hughson in the Sea Cadet Hall and an afloat section came into being at Ardrishaig under Colonel Colquhoun. South of the Border an important new Unit to strengthen the Tyne area was opened at Sunderland. Orkney became important as an assembly anchorage and as a North Sea Oil terminal, so Unit numbers were increased at Weyland Camp, the TA Centre at Kirkwall. Across the Pentland Firth, the Thurso Unit acquired the old Fisherman's Rest which for many years had remained a desolate and unwanted building. A dedicated and hard-working unit set to with scrubbing brushes and pails to clean up a century of dirt and dust. They made repairs to falling plaster and to the hole in the roof. They provided tables and chairs. Such was the beginning of this and other Units, struggling to live on a low budget and members doing much of the work themselves.

In 1969 an important advance in training was made with our first Instructional Technique course at the Royal Naval Air Station at Arbroath. Training improvements were achieved steadily throughout the years and until we were firmly committed to the annual exercise programme, were much assisted by annual competitions both inter Unit and inter area. The Gallaher Trophy, the Dewar Trophy, the Yates Cup and various efficiency shields provided not only the chance to improve performance but were pleasant social occasions allowing members to meet to exchange ideas and to share their problems. The Tees and Tyne areas were prominent in these competitions where Lieutenant Commander 'Buck' Taylor had become the Training Officer and Lt Cdr Tom Reed RNR the PNXO for Tees. These North Eastern Coast Units increased in strength and expertise and made a great contribution to both National and NATO exercises in spite of inadequate training centres. An afloat event of significance was the attendance in June of XSV *Pagham* with a Humber Crew at the large NATO Naval Review at Spithead. In July, with the introduction of Area Flag Officers, the Humber Units were transferred to FO Medway's administration but with FOSNI retaining operational command. This proved a workable solution to the boundary problem, although hardly a satisfactory one.

In January 1970 all were saddened by the loss of the Fraserburgh lifeboat and in particular the coxswain, John Stephen, who was the Head of the Fraserburgh Unit; unfortunately the Unit had to close after this tragic accident.

The purpose built Training Centre for the Inverness Unit at Muirtown was opened in February 1971. The ATO, Harry Fowler, retired and was relieved by Lieutenant Commander Neville Kempton. In the Forth Area Major Price turned over the reins of PNXO to Commander Alan Anderson RNR. During this period the 'Scottish System' for Port Headquarter displays was introduced and served the needs of NCS for many years.

In 1972 our first of the new class of XSV was accepted for service. The Fleet Tender, *Loyal Proctor,* steamed north from the Humber in December with a local crew, to be based at the new training centre at Muirtown. Being winter-time, she carried out much of the familiarisation and initial training in the relatively sheltered waters of Loch Ness. New Units were established at Leith and Methil and the Newcastle Unit was re-activated. Later in the year XSV *Birdham* was again involved in fire fighting. This time she embarked the Rothsay Fire Brigade and sailed for the East Kyle of Bute where the tanker *British Valour* was on fire: the fire was contained and doused successfully.

1973 was a busy year and a vintage one for training centres. The Ayr centre was opened by Mr George Younger MP at Prestwick. A month later the new purpose built PHQ and training centre at Scapa Bay for the Kirkwall Unit was opened by the Lord Lieutenant of Orkney—Colonel R A S MacRae. A new jetty adjacent to the training centre at Muirtown was completed, both facilities being put to good use that year during the Around Scotland Power Boat Race and the Blue Water Trophy race to Bergen. The Elgin Unit moved into the old Civil Defence Centre at the Inefield Estate. The Ardrossan Unit moved up the Clyde to the NATO Boom Depot at Fairlie and in Northern Ireland the Bangor Unit separated from the large Belfast Unit and set up on its own.

It became one of the *Loyal Proctor's* tasks to embark survivors from the battleship HMS *Royal Oak* for their annual wreath laying ceremony in Scapa Flow. Also that year, it was the turn of the Inverness Unit to assist a Fire Brigade by providing a communication net during a serious forest-fire at the Seafield estate. The type 626 VHF radio equipment recently acquired proved invaluable to the success of this operation.

On the Clyde Norman Drake, the Training Officer, resigned in 1976 and transferred to the Clyde River Authority; his post was filled by Lieutenant Commander David Potter RN Rtd who became the Group Training Officer for the Clyde and Northern Ireland. At Pitreavie there was a change of Area Training Officers when Lieutenant Commander David Pugh relieved Commander Neville Kempton. Commander Charles Emerson retired January 1977 after 15 valuable years' work which spanned a period of considerable change; much of the credit for the improved standards of efficiency was due to his leadership. He was relieved by Commander Ted Quance, promoted from his post as ATO Plymouth.

1977 was of course the Royal Silver Jubilee Year and in mid-June the Area craft set off for Portsmouth. They were: XSV *Tongham* (PNXO Bob Paterson), XSV *Pagham* (CNX Derek Edminson) and XSV *Loyal Proctor* (SNXO Ian MacDonald). This was a unique occasion that demonstrated clearly the ability of the RNXS to provide an efficient and cost effective contribution to any commitment whether ceremonial or emergency. A new PHQ and training centre Tees was established in Hartlepool and at last this Unit had adequate facilities under the supervision of the new PNXO Eric Goss. In the adjacent area Alastair Cochrane-Dyett took up the reins as PNXO Tyne. The Area Training Officer, David Pugh, left to join the Government Communication Centre at Cheltenham and was eventually replaced by Lieutenant David Gilbertson RN. Lieutenant David Wakefield joined as the new Area Engineer Officer. Louis Newell relieved Bill Craig as PNXO North Scotland. The second of our new fleet tenders, *Supporter,* was accepted for service on the Clyde and in Northern Ireland, replacing the ageing *Tongham.* The third, the *Loyal Volunteer,* replaced the *Pagham* on the east coast. The prefix Loyal was omitted for the *Supporter* after much soul searching. It seemed the best way to remain non-political in this area.

In 1979 a new task for the Forth Units arose, to provide afloat control for the International Dragon sailing events in the Firth. *Loyal Volunteer* played an important role and a request for similar support in 1985 followed. At Lerwick a few volunteers formed the nucleus for the start of a new Unit. The only training space available was one small room in Fort Charlotte. Accommodation shortage in the Shetlands was to delay the blossoming of this Unit for several years. The Elgin Unit, temporarily re-housed with the RAF at Lossiemouth was soon to find a more permanent address in the town. Deterioration of Stranraer Castle required the temporary re-housing of the Unit at the RAF establishment at West Freugh. They were later permanently housed in the town under the Sheriff's eye.

Another trip to Norway. Loyal Proctor *and* Loyal Volunteer *stop overnight at a Norwegian Village—a long way from home.*

Following the lifting of the moratorium on exercise expenditure in 1980—82, it was good once again to be able to arrange interchanges between Units for training and mutual support. The live aspects of local area exercises were always welcomed, and when Headquarters were activated during weekend exercises with active service and RNR participation, members appreciated the opportunity of serving with a professional team.

1982 saw the first of our Fleet Tenders enter the synchrolift covered repair facility in Rosyth, opening a new era for quicker maintenance. In October the Tyne Units organised and participated in a most moving ceremony in the dedication and laying up of the RNXS Ensign at St Nicholas Cathedral in Newcastle.

◀*Another medal—this time to Lieut. Fred Hull from the Flag Officer Scotland and Northern Ireland—Vice Admiral Tom Baird.*

Clear lower deck of the Firth of Forth Units—August 1984—to say farewell to the author.

With the closure of the Medway Command in 1983, the administration of the Humber Units, under their renowned and long-serving PNXO, Captain Arthur Harvey, was transferred back to FOSNI, so ending the uncomfortable split between administration and operations. The *Loyal Helper* also transferred from Chatham to the Humber, later to be based at Rosyth for East Coast training. In November, Units celebrated the 21st birthday of RNXS in their own various and convivial ways.

Without doubt, a climax for Northern Area Units came in March 1984 with a major exercise, "TEAMWORK", at Invergordon. 31 Merchant Ships were involved and local RNXS resources were stretched to the limit. During a busy week, there was much job satisfaction gained in continuous watchkeeping and incessant XSV running and all felt full of confidence afterwards. Later in the year we said farewell to our stalwart Northern Training Officer, Lieutenant Commander Erlend Clouston, after 17 years dedicated service and also to our much respected Administrative Officer, Lieutenant Fred Hull, after 18 years patient and industrious service. These two officers had contributed an enormous amount to the RNXS. We welcomed Lieutenant Peter Clark from the Medway as Fred Hull's relief and recruited two new Group Naval Auxiliary Officers, Lieutenant Commander Mike Bull for North Scotland and Lieutenant Phil Sparkes for the Northern Isles. We bade farewell to PNXO Brian Watt MBE, a pioneer who served for a total of 30 years. Joe Weir was his successor. After the post had been gapped for six months, Commander Ian Smith, RD*, RNR, became the Area Training Officer, Lieutenant Frank Savage became the first GNXO of the Humber Area and at the same time the PNXO, Captain Harvey, retired.

The Blyth Unit moved into an excellent new Training Centre in July 1985, a blessing derived from the re-development of the port. In October of the same year Commander Ted Quance retired and Commander Ian Smith took over as ANXO.

In 1986 Lieutenant Commander Buck Taylor retired after 17 years' invaluable service in the Tyne and Tees Areas. The Aberdeen Unit moved to a new PHQ and training centre in the harbour offices and the Invergordon Unit into a splendid office block vacated by the RAF Marine Branch. Later in 1986 the Stockton Unit, together with RNR Communications Centre, transferred to a new Tees Training Centre and PHQ at Middlesborough, opened on 16 November by the Commander-in-Chief Naval Home Command, Admiral Sir Peter Stanford. At the end of the year, the Shetland Unit moved from their leaky shack at Victoria Pier to modern premises in Charlotte House, Lerwick. This, after some ten years of hoping and searching, now provides an excellent training centre and PHQ. Thus over the years, the accommodation problem has gradually been resolved, although so often it has been a case of two steps forward and one step backward.

A welcome development of recent years has been the series of routine exercises in which RNXS members, together with the Naval Control of Shipping Officers, train as a team at weekends in their own PHQs. This has taken many years to achieve in the FOSNI area, due largely to geography and its time and distance factors, but it has led to a significant improvement in the standards and capabilities of both organisations. Membership in the Area now stands at 1100, with 26 active and one dormant Units. The development of the GNXO scheme has brought added vibrancy and zest to what has always been a remarkably enthusiastic organisation. Older members have noted a much younger element following in their footsteps, who having sampled the unique cameraderie of the service, are staying to provide its lifeblood and giving continued expression to the voluntary spirit, a hallmark of the British people which has not diminished through the years.

Chapter Ten

'Boring' Finance

General Aspects

The Royal Naval Auxiliary Service began on a low cost basis and, as explained in earlier chapters, was given no money (votes) of its own apart from a small contingency fund for incidental expenses. No cost ceilings were set, because the financing was carried through in the margins of the various votes of the parent service. For example, recruitment expenses were paid for from the Naval Personnel recruitment votes and each year a forecast was made and an explanation given on behalf of the RNXS of how much money would be needed for the following year. Provided the forecast accorded with the current policies and was accepted by the vote holder, no further action with the vote was needed and the RNXS went ahead with its intended recruitment. The advantage was that all the accounting was done for the service by MOD staff; the disadvantage was that, in most votes, RNXS costs were not separated so it was often difficult, when challenged, to know exactly how much had been spent on the RNXS. In the case of recruitment, carried out locally by each Unit, the expenses have always been relatively small and even when, due to a moratorium, finance ceased, recruiting continued by word of mouth. Most expenditure soon becomes a matter of priorities and usually the RNXS was at the lower end of these. Nevertheless it was appreciated by Senior Officers and MOD staff that while the service had to prove itself and be cost-effective, the members needed support and encouragement if the Service was to prosper.

Fortunately in the early days, financing was in the hands of capable and experienced Admiralty civil servants, well accustomed to dealing with such extraneous naval activities and who often knew more about the Service's requirements than the permanent officers in it! The RNXS was launched in favourable financial circumstances as "a creature of Admiralty" whose civil servants saw the Service as their responsibility. With the advent of the Ministry of Defence, the change in attitude was imperceptible at first but gathered pace steadily in the 1970s. Financial control and disciplines became more strict and the requirement to feed remote and impersonal computers with information became a chore which gave little in return. Many years were to pass before computers were developed and programmed to provide answers that made sense and it was during this period that the RNXS lost its one piece of financial independence, the RNXS Contingent Fund. This was merged with a miscellaneous naval vote to avoid overlap and duplication, causing little difference administratively but creating a feeling that the RNXS was not competent to run its own fund. Good finance perhaps, but not good for morale.

Self-training at Units had placed a responsibility on the administration for finding stores to equip all the training centres. Those centres which were also Port Headquarters had some advantages because they were entitled to an establishment of stores while the Units had to bid for training equipment on educational votes. An early lesson was learnt from 'the case of the parallel rulers'. Soon after the RNXS was formed and Plotting became an essential feature of exercises and training, there came a demand from Units for the roller type of parallel rulers, superior to the wooden sliding type. On the basis of a minimum of two rulers per Unit, the RNXS requirement was some 160. Naval stores held fewer than 100 for issue for all fleet priorities and purposes and were prepared to release only 10 for the RNXS! It was to take five years before all Units received their rulers.

Capital costs of training craft and permanent stores are borne by the Navy, and the RNXS has had to argue for its needs and allocation like everyone else. However, given the RNXS role and responsibility there was not usually much difficulty in getting agreement in principle; the problem lay in finding the money and getting the priority to buy the craft or equipment.

In many respects, not least in the procurement of funds, it has been of great benefit to the RNXS to keep a foot in the Naval Staff door in the person of the Command Naval Auxiliary Officer. By maintaining a two days per week presence in the MOD, keeping abreast of financial matters and noting trends and shortfalls in expenditure, he has frequently been able to implement RNXS re-equipment programmes bit by bit. At other times redundant equipment has been diverted from its likely destination as scrap to fulfil a useful function in RNXS training: for example, an urgent requirement for 60 teleprinters was satisfied when an establishment overseas closed.

Training Centres

The strength of the RNXS lies in its Unit organisation. Most members on duty, for most of the time, are able to live in their own homes, requiring no expenditure for food or accommodation arrangements but needing somewhere to train. Each Unit now has a Training Centre but in 1962 many Units had to meet and train in hired or borrowed premises in difficult circumstances. With the increasing responsibilities of the Service, money has been expended over the years in bringing the majority of Training Centres up to standard and providing the necessary facilities. The problem in achieving this has been that most Training Centres are used only once or twice a week, some also for perhaps half a dozen weekends and for an annual exercise for up to three weeks. When in competition with other Government bodies requiring full-time use of premises, the RNXS tended to slip down the order of priority.

Ideally, a Training Centre needs to be located in a large building that is in regular use, has the necessary domestic facilities and is cleaned and heated. Where a Centre has to double as a Port Headquarters, however, siting is of considerable importance. Seldom is the ideal achieved: frequently the costs are too high and the search for suitable premises may be never ending. The process is cumbersome and slow, starting with the Area Office's proposal to the Area Flag Officer, thence to the MOD, to the Property Services Agency, to the local Land Agent for a report and back through the chain again. This often means someone else gets in first before even the finance can be arranged. In the early days those Units accommodated in Admiralty property were well off and the Admiralty owned land and property in many of the larger ports. Often these had been leased to Port Authorities, Customs or other agencies and were only recoverable for emergencies or war. When the Admiralty disappeared, its property was absorbed into Defence Lands, thereafter into the Property Services Agency and "rights" to properties not actually occupied by the Navy were lost. There was the bonus, in theory at least, that the range of Government properties available under the PSA was much wider, although a small organisation did not stand much of a chance with the better properties. Nevertheless, with patience and local initiatives by the RNXS and understanding and help from the PSA agents, improved Training Centres were slowly produced but not before members had suffered discomfort without complaint and so often had made the best out of the worst by their own efforts. It is doubtful whether naval and reserve personnel closing up at these Training Centres and Port Headquarters for exercises appreciated just how much had been achieved.

During the 1970s there was a great contraction and closing of Naval and Admiralty establishments throughout the United Kingdom. Many of these had been props, supporters and sources of advice to their local RNXS Units and they were sadly missed, several Units feeling very much alone having also lost their Training Centres.

Today with fewer RNXS Units, better administration and closer co-operation with the Royal Naval Reserve over the 'Navy House Concept', Training Centres are on firmer foundations. The Navy House concept attempts to bring together under one roof all Naval-related organisations in a locality, each having its own offices and, where appropriate, all sharing common facilities, enabling the building to be fully used and saving on overheads such as domestic services, heating, cleaning and security.

The concept was recommended by Rear Admiral Mitchell in 1974, but human nature, lack of urgency over property matters and lack of agreement over finance have all taken their toll in the thirteen years since. "Banging heads together" had its attractions for management self-esteem but proved counter-productive so often over accommodation matters.

Emoluments

The RNXS permanent staff are linked to Civil Service Executive Grades for pay, are administered as established civil servants and are recruited through Civil Service Commissioners' open competitions. Although open to anyone with the right qualifications, the majority of the officers are ex-RN with a sprinkling of ex-RNR or former Merchant Navy officers. The hours worked are long and frequently unsociable but the compensations are plentiful and the job satisfaction considerable through working in a voluntary and enthusiastic environment. So much is this so that it tends to become an all-absorbing way of life, potentially damaging to domestic harmony unless a proper balance is maintained. The part-time staff, the Port Naval Auxiliary Officers, are deemed to work for 10½ hours per week in their RNXS capacity, in addition to their normal civilian occupation, and are paid at one quarter the top Executive Officer grade salary for so doing. Seldom over the years have PNXOs restricted their RNXS hours to those for which they are paid; on the contrary, most work for much longer in aggregate and many have devoted their 'spare time lives' to the Service. Theirs is indeed a proud history.

Volunteer members would be paid on call out; during peacetime they receive only allowances. Allowances fall into three categories: for the training or duty undertaken, travelling on duty away from the Training Centre and for the upkeep of uniform on loan. From the earliest days, the principle applied was that no volunteer should be out of pocket for any duty performed, not an easy matter to arrange nationally due to local variations. In any case, for most RNXS members volunteering their services, financial gain is not a factor. Furthermore, these allowances have to be claimed after the event, frequently a time-consuming task and, following a major exercise with a large number of claims to be checked and processed, members sometimes do not receive their dues for six weeks or more. Administrative Officers need to work exclusively for several days checking and authenticating the claims and while efforts have been made to resolve the delay problem, it is a stubborn one and only little improved. While RNXS Units continue to be widely dispersed throughout the Kingdom, MOD financial services have contracted and centralised away from establishments to bases and now to MOD itself; utilising computers to provide speed and savings on staff but eliminating the personal touch of earlier, more flexible days. However, the Defence Accounts staff who only recently have assumed responsibility for RNXS allowances are proving efficient and responsive to the special factors which apply and delays are, perhaps, slowly becoming less.

Advances on allowances prior to exercises may be claimed but it is not always possible to gauge sufficiently far in advance how much expenditure might be incurred, nor by whom. Advances have worked more satisfactorily with scheduled courses, where expenditure can be predetermined.

Members are eligible for travel allowances and subsistence at the same rates and on the same basis as civil servants, so there has been no difficulty in keeping these up-to-date. As unpaid volunteers, members are also eligible, like servicemen but unlike civil servants, for 'home to duty' travelling allowances for weekly attendance at their Training Centre. This being a local matter with many variations, the control of home to duty allowance was delegated to Captain RNXS.

Originally, allowances for clothing included uniform cleaning and repair allowance (1962 rate 15 shillings a year); uniform alteration allowance and shoe repair allowance, both at 75p a year until they were abolished in 1979. Only the cleaning and repair allowance remains and is now £5.35 per annum. No other allowance is paid for kit upkeep; members are issued with uniform 'on loan' and may exchange worn articles for new on a one for one basis, subject to fair wear and tear.

The training or duty allowances payable are a more interesting study as they reflect the dilemma of how to recompense Volunteers for their services. There are three: Training Expense Allowance (TEA), Course and Exercise Allowance (CEA) and Supplementary Loss of Earnings Allowance (SLOEA). From the formation of the Minewatching Service, Training Expense Allowance was paid for short periods of up to 10 hours: that is, mainly for the weekly sessions of evening training carried on throughout the year. Its purpose was to compensate Volunteers for inconvenience and to enable them to buy a meal or some refreshment, because in many cases they did not have the time to get home for a meal before the training session started. A similar allowance is payable to the Territorials and RNR. The RNXS members were given the basic rate (1962—25p for 2—5 hours; 50p for 5—10 hours) and the allowance has kept more or less in line with other costs through the years. Today there is just one rate for all: £2.35p for 2 up to 8 hours. Probably not quite so often these days but formerly at some Units, members would donate their training expense allowance to the Unit fund and thus save corporately to pay for Unit entertainment and social events.

Course and Exercise Allowance was introduced for members who undertook courses or exercises lasting 48 hours or more. Why 48 hours? Because at that time volunteer reserve personnel were paid only for periods of training or duty in excess of 48 hours; any lesser periods being covered by TEA. It was thought improper for the RNXS to enjoy a shorter qualifying period, so 48 hours it was—and still, with minor exceptions, remains, although today RNR personnel are paid from day one and proportionately for periods of less than a day. The Allowance was envisaged as pocket money to cover incidental and miscellaneous expenses, standing at 7 shillings and 6 pence per day from 1962 for some ten years or so, when it was increased to £2; currently it is £12 per day. The details of eligibility were left for the CRNXS to determine. In 1972 the Treasury, fearful that Course and Exercise Allowance should be drawn for long periods and constitute in effect "employment", set a limit of 21 days per member per year. With exercise participation gathering pace at that time, an appreciable number of members reached the limit and though willing to take part in further exercises, were not eligible for the allowance. More importantly perhaps, they were not covered for National Insurance benefits, for which the MOD paid the contributions of those in receipt of Course and Exercise Allowance. The ruling was hotly debated at the time because, while no one wished to prevent Volunteers from taking part in courses and exercises, it was felt that equal opportunities ought to be available throughout the Service to spread the training benefit and to avoid the charge that "the same faces appear at every exercise". However, a year or so later the limit was quietly extended to 28 days and provision made, on application, to allow aggregation over a two—year period.

Supplementary Loss of Earnings Allowance was introduced in 1962 to assist members who suffered a loss in earnings through attending exercises. It was designated 'supplementary' to recognise that losses could not be met in full and the allowance was for ameliorating the loss. It was intended primarily to help those who worked in shifts or who were paid by the day; it was not aimed at the majority who worked normal hours and had their evenings and week-ends free, nor the self-employed who were mostly able to arrange their own programmes. In fact the RNXS had been offered a Bounty similar to that paid to the Royal Observer Corps but the management at the time were against bounties as being difficult to administer fairly and likely to attract the wrong type of Volunteer—"the bounty hunter"—that is, the member who makes sure he qualifies for the bounty but does little else! SLOEA was thus substituted, starting with a maximum of 7 shillings and 6 pence per day. Within two years this had risen to £1-12s-6d and the Treasury started to take a keen interest! As "earnings" it was made a taxable allowance; the claimant was required to obtain his employer's signature to a form stating that he had not been paid for the period of training and the allowance was linked firmly to CEA except that, unlike CEA, it could not be claimed for Sundays. These measures were challenged without result and it was some time later that the reason for them became clear. The Treasury, looking for a precedent for this new and unusual allowance, had decided that another Loss of Earnings Allowance, unknown in MOD but payable to civilians required for duty as jurymen or for other Crown

Service, was very similar and had applied the same rules to the RNXS allowance. The allowance did have some effect between 1966 and 1975 in helping those attending exercises, then the employment situation became very difficult; members were reluctant to go to their employers for a signature which might have the effect of taking them off the payroll. As for the employers, they were reluctant to involve themselves in the administrative turmoil of adjusting for pensions and social security payments. So the allowance has been comparatively little used, thereby saving a great deal of money (£18,000 in 1964 alone), which might have gone to those who lost income through volunteering. Today the 'self-employed' are the main beneficiaries: not what the originators had expected. Possibly, now that Unit supervision has been strengthened through the Group Naval Auxiliary Officer scheme, the bounty could be reconsidered even though there are disadvantages.

Until 1972 the Pay on call out for RNXS members was linked directly to HM Coastguard but in that year the Coastguards underwent a considerable re-construction, so that the link was no longer appropriate. There followed a period of financial restraint when no major pay reviews would be considered but, eventually, unsatisfactory and tentative scales of pay for use in emergency were negotiated. In 1980 an effort was made to conduct a pay search review; it did not succeed but what did emerge from the discussion was that the question of pay on call out could not be settled in isolation from matters of status and commitment now that the Service had an important role in war plans.

As discussed in Chapter Twelve, these and related factors are being considered currently and a set of proposals have been formulated which, if approved by the Admiralty Board, will enable this long-standing anomaly to be resolved.

Chapter Eleven

Uniforms and Badges

All shapes and sizes—and a wide range of uniforms too!

Uniforms and the badges and insignia which go with them are as important to auxiliaries as they are to active service personnel. For the Volunteer training part-time for short periods in peace, uniform provides a visual link and bond with the parent service and demonstrates the essential element of 'belonging'. Badges and insignia are particularly significant for RNXS members since they denote the qualifications and progress of individuals whose service is rewarded chiefly by a sense of achievement and not by financial gain.

The story of the supply of uniform to the RNXS is marked by struggles against financial stringency, by the achievement of minor improvements year by year and by many problems over sizes, fit and distribution. Not much different from the average shopper in the United Kingdom, the cynic might say! Such is the resilience and patience of the average RNXS member that, with much self help and with few exceptions members have always achieved a high standard of turn-out. There was even a period of nearly a year during the late 1970s when no kit at all could be supplied and yet, with good family housekeeping by the Units, no drop in standard of dress was apparent except that some new entries were obliged to make do without various articles until normal supply was resumed.

When the Minewatchers first started in 1952 they wore civilian clothing with a Minewatching armband but it was soon realised that the exposed nature of their job called for

The famous ''bull's wool'' uniform, warm but tickly.

suitable tough, warm and protective clothing. The first articles issued on loan were berets, greatcoats and boots, with shoes as an option for ladies. There was a long discussion on which uniform should be adopted and the Admiralty finally directed that existing lines of Service clothing were to be utilised and only an essential working kit would be authorised. The battle-dress tunic style as worn at the time by the Royal Observer Corps and Civil Defence was adopted. This translated in naval terms to the working dress made of sailor's serge cloth ('Bull's Wool' in slang), of which there were large stocks worn chiefly by the Fleet Air Arm. This uniform looked reasonable on the younger members with figures fed on a wartime diet and corresponding to the standard sizes available. There were, however, severe problems in altering issues to fit the mature fuller figure, particularly the fair sex where sometimes the chest and waist measurements were difficult to co-ordinate in one issue. A request for a made to measure garment was considered an insult by the clothing department and seemed to require Treasury approval on each occasion. Nevertheless, within a year an alteration allowance for individuals was approved and as the years went by sizing gradually improved but never quite caught up with the expansive diet and growth of some of the older generation.

Members greatly appreciated the warmth and protection afforded by their uniforms and even the boots were quite popular, although the procedure for fitting a new pair by forcing them on and then standing in hot water for half an hour while the leather shaped to the feet seems incongruous today. Blue cotton twill shirts (No 8) were supplied with the uniform and for formal occasions black ties were worn. The collars of No 8 shirts were not designed for wear with ties and unless great care was taken, a ''Just William'' appearance resulted.

To supplement their basic kit, members were entitled to buy certain items such as gloves, extra shirts and shoes from service clothing stores (slops) ''on repayment'' at discount prices. The word repayment, meaning to pay back, applies particularly to regular personnel using their clothing allowances to purchase items of clothing for themselves. It was represented that Volunteers, who received neither pay nor clothing allowance, should be granted special discount prices for certain items but this did not find favour.

Proficiency badges displaying the Minewatching sight were awarded to Qualified, Leading and Chief Minewatchers and worn on the right arm. A Chief appointed as Head of a Unit wore a large Crown on his left arm. Officers wore the same basic uniform as members; their badges of rank were made of silver braid and worn as shoulder straps. Silver braid was authorised to indicate that the officers were not commissioned and that their service had a civilian as opposed to a Royal Navy basis. For the Minewatching badge and crest, needed urgently for wear on the beret, ''the splash of a large air-dropped mine on impact with the sea'' was chosen, designed and approval obtained mainly by Commander John Forbes, who had sponsored the Minewatchers in the Admiralty. On formation, the Royal Naval Auxiliary Service adopted the badge, with an amended inscription and it continues today.

The changed role of the newly formed RNXS reduced the requirement for robust outdoor clothing and the need for an improved appearance when working alongside other services became apparent. As ever, money was short, so improvements were spread over the years. Seagoing kit for the Afloat branches and a better fitting blue shirt took much of the 1963-64 allocation, so major improvements to the uniform had to be postponed. A further desperate effort to improve the ladies' uniform fit was made; however, it seemed that the combination of thick wool serge and the cut and style of the working dress blouse could never suit the mature female figure. It is known that a number of valuable recruits to the Service were lost because of this problem.

Clothing is issued on personal loan to serving members, to be returned on leaving. Worn or damaged items are replaced on a one-for-one basis, subject to rules and guidance on normal wear and accidental loss or damage. Until the early 1970s, kit returned was sent back to the clothing depots for re-issue or shredding for rags if unsuitable. Modernisation of the supply system and streamlining of staff then made the central re-issue

The "new look" uniform, a smoother and smarter cut.

arrangements uneconomical, so Units with storage space were invited to hold re-usable items of returned clothing for local issue in emergencies. This proved a useful "housekeeping" and cost-saving arrangement.

On formation, RNXS badges for the beret and shoulder insignia were supplied promptly and an RNXS pocket badge in yellow lettering with a grey cloth backing to brighten the rather flat appearance of the blouse was tried, the yellow and grey representing gold and silver. Although the idea was liked, the badge, not surprisingly, was not and after two years a more attractive design in white on a blue background was substituted. While the new training organisation for the RNXS was being developed, the basic Minewatching badges were continued for qualifications as an interim measure, with the branch being indicated by a single letter P for Plotting, C for Communications and so forth. As training and qualification developed along Naval lines, it was felt undesirable to introduce special branch badges for the RNXS, so approval was given to use and adapt existing RN badges. This was undoubtedly an aid to understanding and integration at local headquarters.

By the late 1960s the higher profile of the RNXS on duty at Headquarters, recruiting, parading and visiting foreign ports, demanded better uniform. Complaints such as "Nobody takes us seriously" or "This uniform is compulsory for the inmates of HM Prisons" were voiced. Nevertheless, not until February 1971 was approval given for a new working dress uniform made from diagonal serge and with much improved cut, in a few standard sizes and with clear instructions for 'made to measure' should this be necessary. By this time there were only a few other Services using the working dress, so the contract was a relatively small one and therefore not popular with the garment manufacturers and the unit costs were high for a low-cost organisation such as the RNXS. Although it took nearly three years to complete outfitting, the new uniforms were a great improvement, providing a more sleek and efficient image. Pride in the Service increased and the demand grew for RNXS members to attend such functions as the Royal Tournament, the Boat Show, the Festival of Remembrance in London and similar public events at other towns and cities. The uniform was worn for the first formal occasion by RNXS members lining the route at Caernarvon for the induction of Prince Charles as the Prince of Wales.

In the early seventies the Royal Navy introduced the Heavy Wool Jersey (HWJ), which soon gained acceptance as a comfortable and practical everyday working rig; envious RNXS eyes were cast towards it. With a new working rig recently agreed it was difficult to get further approval to add the HWJ to the official kit list. Eventually, approval was given in 1975 for the optional wear by members of the HWJ bought on "repayment". It proved popular with both men and women, although the Wrens' version of the HWJ did not appear until some three years later. The only disadvantage of the HWJ for everyday use appeared to be that it was unsuitable for sewn on badges because they shrank in the wash, so the garment could not be worn on representational occasions.

As explained in Chapter Six, the privilege of wearing full Naval uniform, granted to the RNXS in 1979, was a watershed for the image of the Service. Members at last felt that they were accepted by their parent Service and that the 'Dad's Navy' gibe would be a thing of the past (it had long since ceased to be appropriate). The change was finally brought about by a combination of circumstances: the gentle but persistent and ultimately persuasive arguments of the senior RNXS staff, the growing importance and competence of the members on duty and on representational display and by no means least, the increasing costs of manufacturing the separate, and now unique, diagonal blue serge battledress style blouses. The climate was all the more favourable because, to their

The Old Minewatching Badges

PROFICIENCY BADGES

Qualified Minewatcher

Leading Minewatcher

Chief Minewatcher

Chief Minewatcher Instructor

RNXS BRANCH BADGES AND GRADES

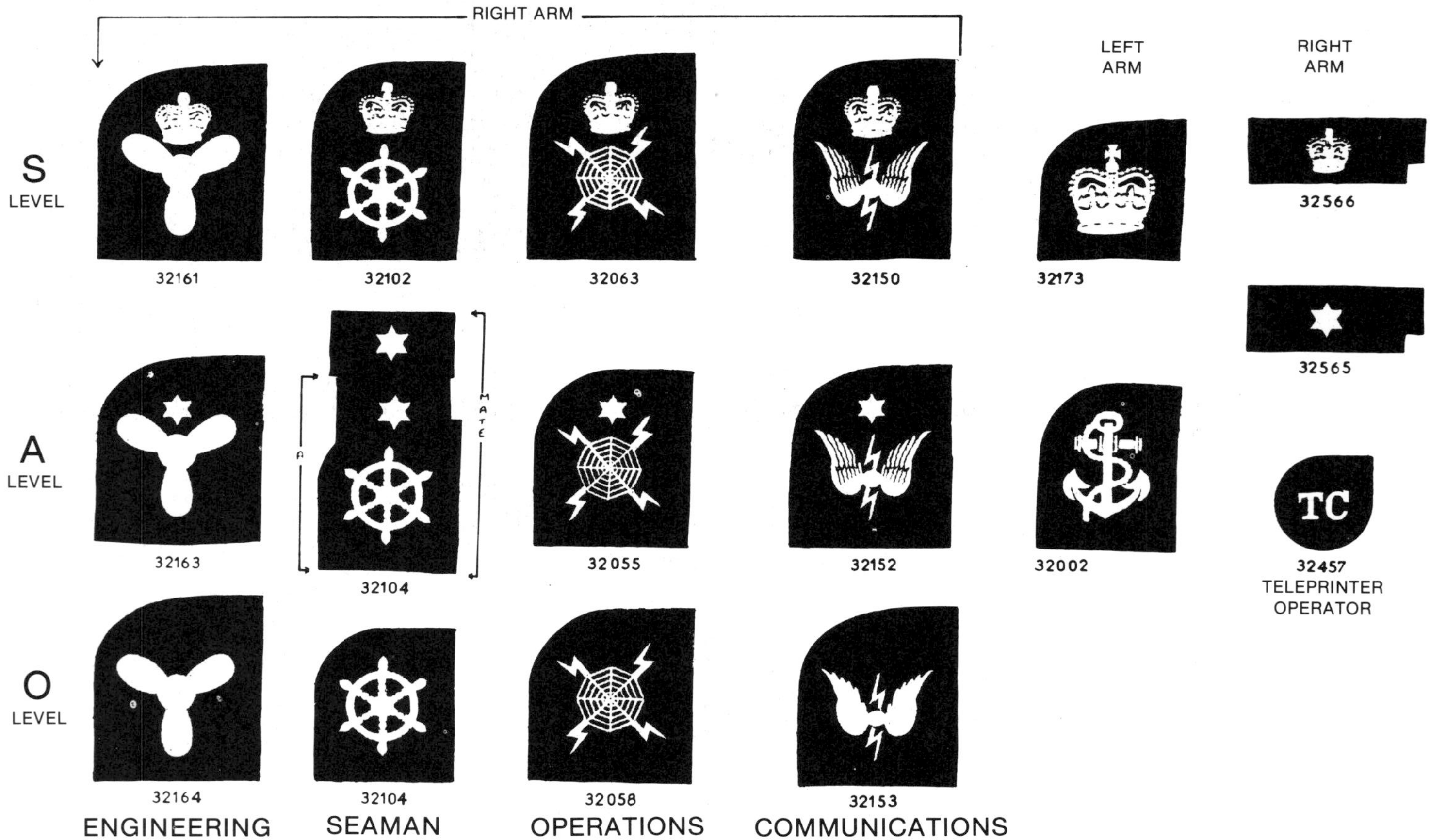

credit, the majority of Volunteers had purchased their own 'woolly-pullies' and thus acquired the optional working dress at no expense to the Crown. There were some who regretted the passing of a common uniform worn by all from new recruit to CNXO; others were disappointed that headgear formed no part of the change. The beret, disliked by some, remained the one common item of uniform for all members.

The original reasons for the beret's retention were its practicability and the high cost of replacement by caps and tricorn hats. It was realised that the beret was a less than perfect complement to the new uniform, particularly for ceremonial occasions, nevertheless it was decided to postpone further consideration until everyone had a new uniform, an interval of almost three years. Upon reconsideration, the costs were deemed still too high and recently a compromise proposal was rejected and the matter shelved for a further period of reflection.

In 1981 Units were invited to submit designs for a new RNXS crest which would better reflect the responsibilities of the Service, so changed since minewatching days, and be suitable as a badge for all purposes including use on the ubiquitous beret. Several interesting designs were conceived, some of which were highly complicated and none considered entirely appropriate. Historical research also proved abortive and was abandoned, the intention at that time being to reconsider the matter later, perhaps in the RNXS Silver Jubilee year. The present view is that the historical link with the Royal Naval Minewatching Service is well worth preserving and that unless some uniquely appropriate alternative comes to light, the 35 years continuity symbolised by the 'Exploding Blancmange' or, as some will have it, the 'Hairy Pudding', will be maintained into the foreseeable future.

Today's 'favourite uniform' the Woolly Pully.

Chapter Twelve

From 1984 to the Future

XSV Example—*''new look'' craft for an increasingly professional Service.*

For the RNXS, 1984 was the year in which significant changes, some of which were described in Chapter 6, began to take effect. Probably the most far-reaching development lay in the decision that the RNXS should take an important share of the Defence of Ports and Anchorages in the United Kingdom, particularly afloat. This required the recruitment and training of a much larger afloat branch and the allocation of faster modern craft. Four P2000 Coastal Training Craft were ordered from Watercraft Ltd to replace *Shipham* and *Portisham* as a first step in this process: two were completed and delivered (*Example* and *Explorer*) but, frustratingly, the firm went into liquidation in 1986 and the remaining two (*Express* and *Exploit*), are awaiting completion. In August 1987, Vosper Thorneycroft was awarded the contract for completion of *Express* (expected early in 1988) and there is good reason to believe that *Exploit* will follow later in the year.

A task for the RNXS in defence of ports had existed from the earliest days because of their commitment to man Port Headquarters. Hitherto, members had given the task little considertion since it had been understood to be, in the Military sense, essentially passive in nature. Now however, the re-assessment of the threat from attack by special forces supported by pre-positioned agents fomenting internal subversion sparked off a new look into the defence of ports, leading to a reappraisal of the requirements for personnel and equipment. Most members welcomed the development but a few expressed concern that the Service, unarmed and, as they believed, non-combatant, was in danger of becoming too closely involved with 'active' defence. It became apparent, for a variety of reasons, that the RNXS constitution needed overhauling. The original constitution stated that the aim of the RNXS was ''..... to provide personnel ..... thereby relieving regular naval personnel for combat duties elsewhere''. To an increasing extent, the nuclear stalemate, the flexible response strategy, the vastly increased threat posed by Soviet Maritime power and the radical advances in weapon technology have combined to render the differences between combatant and non-combatant much less clear-cut. In the history of conflict, until yesterday, a combatant was one who fought hand to hand and carried a weapon. Today a combatant is as likely to monitor an electronic screen many miles from the scene of action. Nevertheless, in the strict sense, the RNXS remains an unarmed Service.

In the changed atmosphere of warfare, or potential warfare what counts is not any fine distinction which may be drawn between 'combatant' and 'non-combatant'—it is ever more true that "he who is not for us, is agin us". Nowadays those who are 'for us' must be ready, capable and reliable. Readiness depends upon recruitment and organisation; capability upon training and experience. Reliability stems from the individual and the moral or material imperatives which cause him to continue to give of his best whatever the dangers and difficulties. Without changing the peacetime, volunteer ethic of the Service, the RNXS staff and those concerned in the Ministry are attempting to adjust the wartime constitution and status of the RNXS to meet those modern requirements in a manner acceptable to the membership.

In the light of the above factors and others such as, the concern of some over the absence of safeguards for jobs and families of members in the event of war, the military command anomalies inherent in the interaction between a 'civilian' and a 'Service' force, and the lack of a realistic wartime pay structure, the RNXS constitution and status have been scrutinised in the context of war. As a result, Ministry of Defence proposals, modified to take account of Naval Home Command and RNXS perceptions, have been put to the membership prior to submission to the Admiralty Board, and 97% of members have indicated their agreement for the proposals to go forward for approval. The two principles which remain paramount are that the RNXS volunteer civilian status should remain in peacetime and that any changes planned for wartime should have the support of the majority of the current membership.

Also in 1984, as described in Chapter 6, the Group Naval Auxiliary Officer (GNXO) scheme began to take effect in the Scottish Area, and was soon followed by appointments in the Portsmouth Area. By summer 1988 new appointments should be complete and the GNXO will have replaced the GTO in all areas. The scheme is assessed as successful and has recently been endorsed by the Director of Naval Manpower Audit. The advantages of having more permanent staff officers based at ports and in direct contact with members is already bearing fruit, but there remain several teething troubles to be solved. In keeping with their new responsibilities, members should benefit from regular in touch briefings, easier access to training, qualifying and general information and a more consistent standard of administration. With the change completed there will be a total of fifteen GNXOs looking after from three to six Units each, while three old style PNXOs will remain to supervise the administration and help with the training of Units under the direct control of their Area Training Officer.

Twenty or so new volunteer unpaid officer posts, christened 'Staff PNXOs' (SPNXO) to differentiate them from the old style part-time PNXO, are in the process of creation. Those in post have already proved their worth, bringing invaluable local knowledge, experience and a 'volunteer viewpoint' to blend with their GNXOs' authority and responsibility. Whilst the primary task of most SPNXOs lies in the PHQ, others assist with training ashore or afloat depending on local requirements and individual skills.

The administration of so many scattered and varied

Prince Michael inspects the parade at the RNXS Review at Whale Island—June 1987.

Units has never been an easy matter and Units nearest naval bases have tended to fare the best. The supply of uniform, stores and equipment is based on naval procedures and methods. Stores accounting, handling and supply have been increasingly centralised and computerised over recent years, with a marked reduction in the number of civilian supply staff. Consequently the RNXS staff and members have had to become more deeply involved with the process in compensation. The payment of allowances and expenses has been 'nationally' centralised on computer and there is now little margin for error; if members don't get their claim form right, they risk not getting the money! These developments have thrown

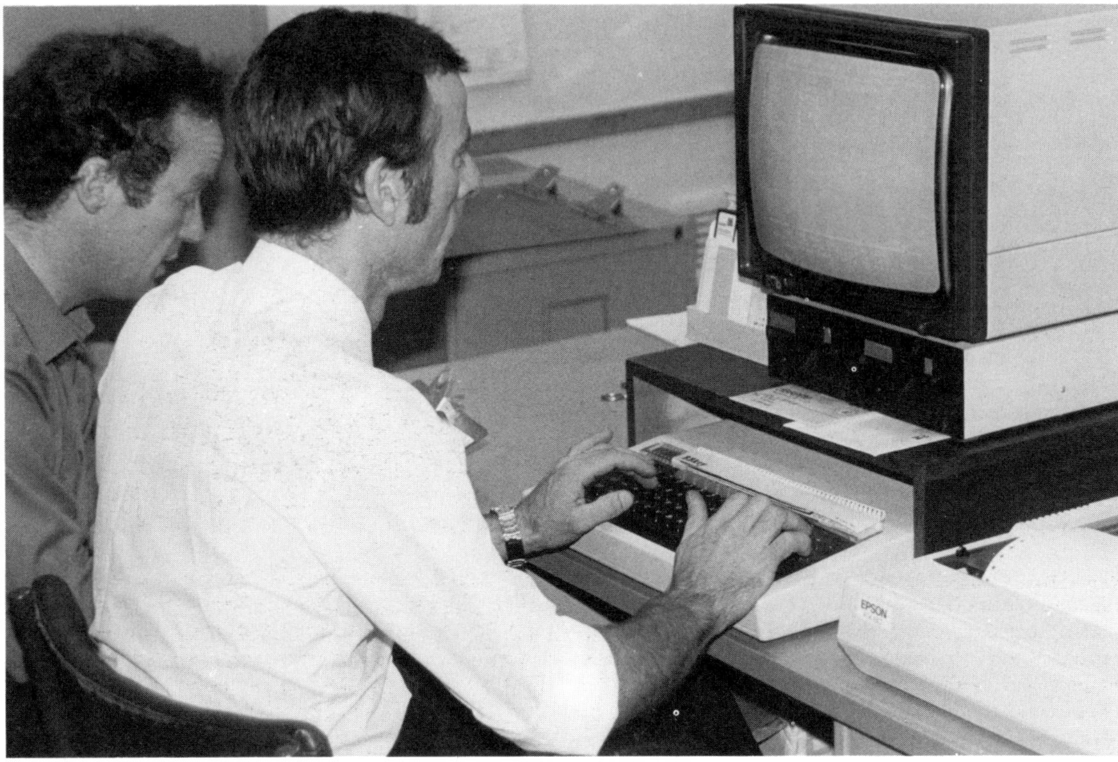

Computers are now in use in Port Headquarters (here at Hull). A far cry from the equipment seen on page 5.

Merchant Shipping in UK waters will continue to see RNXS vessels arriving alongside in the years ahead.

an additional load on Area Administration in checking and initiating action. The process, instead of becoming quick, modern and efficient, at times seemed likely to become strangled with bureaucracy. Happily, however, the manipulators of the Bath computers are proving to be helpful and sympathetic and the dreaded delays in payment have not materialised.

The overload of Area Administrative Staff is an increasing problem and in an era of severe financial stringency and unyielding manpower ceilings, is proving difficult to solve. In the past year the Scottish Area Office has been engaged in trials with data processing equipment and the intention is to install similar equipment at Portsmouth and Plymouth once the system is proven. This should go some way towards relieving Area Staff, particularly the Administrative Officers, from the more tedious repetitive tasks, but it seems unlikely to be the complete answer to the problem and other aspects will need further study.

While these big changes take place, the regular round of training and annual exercises continues unabated and in some areas augmented. In striving to maintain their standards members find plenty of interest and much to challenge their capabilities; the benefit for the nation is their readiness, reliability and growing self-confidence. So in 1987, the RNXS after twenty five years in being, has established itself as a solid support service for the Royal Navy. It is no longer a cheap 'come in handy' organisation but an economical, mature and reliable service that has proved capable and proficient at the tasks assigned. It was extremely proud in its Silver Jubilee Year that His Royal Highness Prince Michael of Kent became the first Patron of the Service.

As this history was being written, further history was created as the RNXS celebrated, at local and national level, its Silver Jubilee. The Annual Church Service on 17 May 1987 at Allhallows-Next-The-Tower was swelled by the attendance of fifty members each from the Plymouth Area and Scotland and Northern Ireland, and honoured by the attendance of the First Sea Lord, Admiral Sir William Staveley, GCB, ADC. A review ashore at Excellent (Whale Island) and afloat in Portsmouth Harbour was held on 27 June in the presence of the Patron. Local unit events were held throughout the Autumn, some combined with Trafalgar Night activities and on November 5th—the birthday proper, the staff and their ladies dine with their guests of honour, the Commander-in-Chief Naval Home Command, Admiral Sir John and Lady Woodward in the Wardroom of HMS *Nelson* to mark both twenty five years of service and a highly significant year in the history and evolution of the RNXS.

In looking forward, in every sense, to the next twenty five years, there seems no reason to suppose that the Royal Naval Auxiliary Service will come to play any lesser part in the Service of the Navy and the Nation. Further change and evolution there undoubtedly will be, some already inherent in the patterns of national and naval development, some as yet 'undreamed of' in any philosophy. Despite hopeful signs, the world remains far from committed to peaceful and fruitful co-existence; there is still a 'lot of wickedness and greed about' and it behoves all those who desire peace to maintain their guard. If financial restrictions continue to inhibit the ability of the regular forces to sustain that guard in every respect, then reserves and auxiliaries must continue to fill the gaps and if the gaps grow . . . . What will not change, in that or any longer period, is the good humoured, patient and persevering loyalty of the RNXS Volunteer, the essential brick with which the Service is built. He—and she—deserve every encouragement and respect and it is largely in their honour that this history has been written. They have served their country well.

LOCATIONS OF UNITS

PORTSMOUTH AND MEDWAY AREAS

BEXHILL*
BOURNEMOUTH
CANTERBURY*
CHATHAM*
CHELMSFORD*
CHELSEA
CLACTON-ON-SEA
DOVER*
EASTBOURNE
FELIXSTOWE*
GATEFORTH
GRAVESEND*
GREAT YARMOUTH*
GREENWICH*
HARWICH*
HASTINGS
HERNE BAY
HOVE
IPSWICH*
ISLE OF WIGHT*
KINGSTON ON THAMES
LITTLEHAMPTON*
LONDON*
LOWESTOFT*
MALDON
MARGATE*
NEW FOREST (BOLDRE)
NEWHAVEN
PITSEA
POOLE*
PORTLAND*
PORTSMOUTH*
SHEERNESS*
SHOREHAM*
SOUTHAMPTON*
SOUTHEND*
WHITSTABLE

PLYMOUTH AREA

APPLEDORE
BARNSTAPLE*
BARROW-IN-FURNESS
BARRY
BIRKENHEAD (LIVERPOOL)*
BRISTOL*
BRIXHAM (TORBAY)*
BROUGHTON MOOR*
BURNHAM-ON-SEA
CARDIFF
CLEVEDON
CONWY*
DARTMOUTH
DOUGLAS, I.O.M.*
EXETER*
EXMOUTH
FALMOUTH*
FISHGUARD
FLEETWOOD*
FOWEY*
GLOUCESTER*
GUERNSEY*
HOLYHEAD*
ILFRACOMBE
JERSEY*
KINGSBRIDGE
LLANDUDNO
LYMPSTONE
MENAI BRIDGE*
MILFORD HAVEN
NEWPORT*
PEMBROKE DOCK*
PENZANCE*
PLYMOUTH*
PRESTON*
SALCOMBE
SHARPNESS
SULLY*
SWANSEA*
TEIGNMOUTH
TORQUAY
WHITEHAVEN
WORKINGTON

* current locations

SCOTLAND AND NORTHERN IRELAND AREAS

ABERDEEN*
ARDRISHAIG
ARDROSSAN
AYR*
BANGOR*
BELFAST*
BIRSAY
BLYTH*
BUCKIE
CAMPBELTOWN
DARLINGTON
DUNDEE*
DUNOON
ELGIN*
FAIRLIE
HARTLEPOOL*
FORT WILLIAM
FRASERBURGH
GLASGOW*
GOOLE
GRANTON (LEITH)*
GREENOCK*
GRIMSBY*
HULL*
INVERGORDON*
INVERNESS*
KIRKWALL (ORKNEY)*
LARNE
LERWICK (SHETLAND)*
LONDONDERRY (Dormant)
LOSSIEMOUTH
METHIL
MIDDLESBROUGH (TEESPORT)*
NEWCASTLE-UPON-TYNE*
NORTH SHIELDS
OBAN
ROSYTH*
STOCKTON
STORNOWAY*
STRANRAER*
SUNDERLAND*
THORNE*
THURSO*
WHITBY*
WHITLEY BAY

* current locations

VESSELS OF THE RNXS

MOTOR FISHING VESSELS (length as indicated)

45ft	61ft 6ins	
657	65	1031
658	80	1068
671	93	1257
637	105	
773	110	

INSHORE MINESWEEPERS

(length 33M, 120 tons displacement)

BIRDHAM
ODIHAM
PAGHAM
*PORTISHAM**
PUTTENHAM
SANDRINGHAM
SHIPHAM
THAKEHAM
TONGHAM
WOLDINGHAM

FLEET TENDERS TYPE C

(length 23M, 115 gross tonnage)

Batch I—*LOYAL FACTOR (HMS ALERT)*
LOYAL GOVERNOR (HMS VIGILANT)
Batch II—*LOYAL CHANCELLOR**
*LOYAL MODERATOR**
*LOYAL PROCTOR**
Batch III—*LOYAL HELPER**
*LOYAL MEDIATOR**
*LOYAL VOLUNTEER**
*LOYAL WATCHER**
*SUPPORTER**

P.2000 PATROL CRAFT (length 20M)

EXAMPLE *
EXPLORER *
EXPRESS Building
EXPLOIT Building

* In service

ROLL OF PERMANENT, PART TIME AND VOLUNTEER AREA STAFFS

PORTSMOUTH AND MEDWAY AREAS

AREA NAVAL AUXILIARY OFFICERS

D W E Baker
D J Farquharson*
J C Ogle
C L Sutton

ADMINISTRATIVE OFFICERS

F W Allerston*
A T Carter
P A Clark
A J Hotham
A G Williams

ENGINEER OFFICERS

J C Kinchin
J E Porter*
M M Trott

TRAINING AND GROUP NAVAL AUXILIARY OFFICERS

D W E Baker
C F Beddoe
P L Brooke*
G J Claydon*
G C Davey
A O Foden
R V Harding*
W C H Harding
S D Jeans
J A de M Leathes
V W Pitcher*

PORT NAVAL AUXILIARY OFFICERS

P W Appleton
A Bates
W Biggs
C E Coult
G H Daniels
D S E Evans
A J Feltham
D Gell
A J Harvey
G Joly
A H Martin
A W Patten
G Ransome
D W H Simmonds
T B Stableford
G A Stratton*
H J T Timewell*
R H White Atkins
S G Willey

STAFF PORT NAVAL AUXILIARY OFFICERS

A J Bates*
D S E Evans*
A W Patten*

*In post (1988)

PLYMOUTH AREA

AREA NAVAL AUXILIARY OFFICERS

R C Beckett
R Gabbett-Mulhallen
B Prideaux*
T S Trick

ADMINISTRATIVE OFFICERS

W H Bethell
E Forbes-Taylor
C F Gray
R J Griffin
W H Metherell
C O Sadler
J W Williams*

ENGINEER OFFICERS

J C Kinchin*
L E Tarry

TRAINING AND GROUP NAVAL AUXILIARY OFFICERS

W H Bethell
D W Bulloch
R B Chandler
G P Curran
C E Gibson*
E C Hill
J E Hodgson
A D King*
D Lermitte
R J Mackrell*
J W H Payne
G C Pidgeon*
R Portchmouth
E Quance
J R Styles
J Thornton
B A Wiltshire*

PORT NAVAL AUXILIARY OFFICERS

R C Beckett
E J M Beely
D W Bulloch
J E Dukes
T H Foden
G W Gallaway
A R Gillingham
M L Hogg*
G D Hunter*
D Keys*
A D King
M E Lee
K MacDonald
C W Powell
C F T Poynder
L Richards
J W Thompson
F W Webster
L Williams

STAFF PORT NAVAL AUXILIARY OFFICERS

D W Bulloch*
J E Dukes*
A R Gillingham*
C F T Poynder*

* In post (1988)

SCOTLAND AND NORTHERN IRELAND AREAS

AREA NAVAL AUXILIARY OFFICERS

C E Emerson
J I D Smith*
E Quance

ADMINISTRATIVE OFFICERS

W M Brown
P A Clark*
J Davidson
J Fawcett
F Hull
R F Lintern
C Powell

ENGINEER OFFICERS

A G Morell
D Wakefield*

TRAINING AND GROUP NAVAL AUXILIARY OFFICERS

D Adey
J Bradnam
G M Bull*
E R S Clouston
N C Drake
J H Fowler
D M M Gilbertson
J I Hughes*
J N Kempton
D K F Potter*
D E Pugh
F Savage*
P J Sparkes*
H A Taylor

PORT NAVAL AUXILIARY OFFICERS

A H B Anderson
A A A Cochrane-Dyett
E G Collier*
W Craig
G Dewer
A Ecclestone
A W Fisher
E V Goss*
W J Grief
A J Harvey
B Hatfield
D G Hedgecock
W S Johnston
C C B Mackenzie
R E Maclean
L W Newell
A Paterson
R Paterson*
I V M Price
T Reed
A N Robertson
G B Sangster
M I Usher
B M Watt
J C Weir

STAFF PORT NAVAL AUXILIARY OFFICERS

B Hatfield*
L W Newell*
J C Weir*
(Mrs) P Woodford*

THE CAPTAINS RNXS

J A MURRAY	NOV.63-DEC.66
D R JOHNSTON	DEC.66-JAN.70
D J FARQUHARSON	JAN.70-FEB.73
P S HICKS-BEACH	FEB.73-MAR.76
G J M ANDREWES	MAR.76-AUG.79
P R D KIMM	AUG.79-NOV.82
M R HARE	NOV.82-NOV.84
J R GRIFFITHS	NOV.84 *

ROLL OF HEADQUARTERS STAFF

COMMAND NAVAL AUXILIARY OFFICERS

Commander R G LIVEING	1962-68
Commander J A MURRAY	1968-84
Captain M R HARE	1984- *

NAVAL AUXILIARY TRAINING OFFICERS

Lieutenant Commander J C OGLE	1962-65
Lieutenant Commander E C HILL	1965-74
Commander D J FARQUHARSON	1974-86
Commander J A de M LEATHES	1986- *

* In post (1988)

DEPUTY/ASSISTANT NAVAL AUXILIARY TRAINING OFFICERS

Mr G W BAILEY
Lieutenant Commander W C H HARDING
Lieutenant Commander S D JEANS
Lieutenant Commander L S SALT *
Lieutenant Commander R V HARDING
Lieutenant G G HORNE *

GLOSSARY

AAO	Area Administrative Officer
AEO	Area Engineering Officer
AFCA	Admiralty Ferry Crews Association
AFO	Area Flag Officer
ANXO	Area Naval Auxiliary Officer
ATO	Area Training Officer
CINCNAVHOME	Commander-in-Chief, Naval Home Command
CNX	Chief Naval Auxiliaryman
CNXO	Command Naval Auxiliary Officer (S = Senior; D = Deputy)
CRNXS	Captain, Royal Naval Auxiliary Service
DEFPA	Defence of Ports and Anchorages
FO	Flag Officer
FOIC	Flag Officer in Charge
FOSNI	Flag Officer, Scotland and Northern Ireland
FT	Fleet Tender
GNXO	Group Naval Auxiliary Officer
GTO	Group Training Officer
HOU	Head of Unit
IMS	Inshore Minesweeper
LNX	Leading Naval Auxiliaryman
MFV	Motor Fishing Vessel
MHQ	Maritime Headquarters
MOD(N)	Ministry of Defence (Navy)
NCSO	Naval Control of Shipping Officer
NCS.Org	Naval Control of Shipping Organisation
NOIC	Naval Officer in Charge
NX	Naval Auxiliaryman
NXTO	Naval Auxiliary Training Officer
PHQ	Port Headquarters
QNX	Qualified Naval Auxiliaryman
RNMWS	Royal Naval Minewatching Service
RNO	Resident Naval Officer
RNXS	Royal Naval Auxiliary Service
XSV	Auxiliary Service Vessel